Computer Graphics and Product Modeling for CAD/CAM

Computer Graphics and Product Modeling for CAD/CAM

S S Pande

Alpha Science International Ltd.
Oxford, U.K.

Computer Graphics and Product Modeling for CAD/CAM
218 pgs. | 149 figs. | 11 tbls.

S S Pande
Professor
Department of Mechanical Engineering
Indian Institute of Technology, Bombay
Powai, Mumbai

Copyright © 2012

ALPHA SCIENCE INTERNATIONAL LTD.
7200 The Quorum, Oxford Business Park North
Garsington Road, Oxford OX4 2JZ, U.K.

www.alphasci.com

Printed from the camera-ready copy provided by the Author.

ISBN 978-1-84265-690-7

Printed in India

To

My Parents

To

My Parents

Preface

Computer graphics is an interesting field which enables the creation of virtual world using digital images. Designers can interactively synthesize 3 D object shapes and transform them dynamically to create animations. Today Computer graphics is widely used in a variety of applications from computer art to satellite imagery and weather forecasting.

Research in Computer Graphics started in late 60s as a sub discipline of Computer Science. In the ensuing decade, automobile and aerospace industries realized the benefits of using Computer Graphics for Product Design and Visualization. Manufacturing industries in early 80s were facing challenges due to shorter product life cycles, frequent design revisions and need for shortest time to market. To meet them, Computer Aided Design and Manufacturing (CAD/CAM) technology was being developed worldwide to introduce automation and flexibility in the product development cycle. Mechanical Engineering curricula traditionally included basic courses on design and manufacturing engineering. A need was thus, increasing felt to train future researchers and engineers on the fundamentals of Computer Graphics and CAD/CAM.

With this objective, during my teaching career at Indian Institute of Technology, Bombay I formulated and introduced a post graduate level (M. Tech) course on *Interactive Computer Graphics for CAD/CAM* in the Mechanical Engineering curriculum in 1986. It spawned activities like contemporary research by several Ph. D and M. Tech students, industry sponsored projects and Continuing education over the years.

This book is primarily written to present the fundamentals of Computer Graphics and Product Modeling techniques for CAD/CAM applications. It is expected to serve as a foundational text book at the senior undergraduate and post graduate level courses in Computer Graphics, Product Modeling and CAD/CAM for Mechanical Engg and Computer Science and Engg students. No specific background is required on the part of the reader except the basic concepts of Linear Algebra and Vector Calculus. The book can also be used as a refresher text by the practicing professionals in industry.

Organized into nine chapters, the book presents mathematical basis for the design of curves and surfaces, Geometric and Projection transformations, techniques for the modeling of 3 D objects and Product Data Exchange Standards.

Chapter 1 introduces basic concepts of Computer Graphics and activities in product development cycle for CAD/CAM/CIM. Role of Product Life Cycle Management (PLM) is highlighted for interoperability and global product development.

Chapter 2 discusses in details, elements of a graphics system in terms of constituent hardware and software. Mathematical basis for viewing pipeline transformations and raster graphics algorithms are presented.

Chapter 3 introduces various types of geometric transformations. Concept of Homogeneous Coordinates is introduced and generalized transformation matrices for various 2 D and 3 D geometric transformations are derived.

Chapter 4 highlights the need to create projections and presents the underlying mathematics using a Virtual camera analogy. Generalized transformation matrices are derived for various types of Parallel (orthographic, axonometric, oblique) and Perspective (single and multi point) projections.

Chapter 5 presents at length, techniques for curve design and their representation in terms of implicit, explicit and parametric forms. Mathematical basis for the design of synthetic curves like Bezier, B Splines and NURBS are presented in details.

Chapter 6 discusses techniques for the representation of surfaces in terms of implicit and parametric functions. Mathematical basis for design of sweep surfaces and bi parametric surface patches (ruled, Coons, Bezier, B Spline, NURBS) are presented with the associated surface properties.

Chapter 7 highlights various issues in geometric modeling of 3 D objects. Wireframe and B Rep modeling techniques are discussed in details, in terms of their topological representations, validity and associated data structures.

Chapter 8 presents at length three object modeling techniques viz. Solid modeling using CSG, Feature based part modeling and Variational geometric modeling. Mathematical basis for object representation, topology/validity and associated data structures are discussed with illustrative examples.

Chapter 9 highlights the need for product data standards for interoperability among heterogeneous CAD systems. Formats for the representation of Product Data for two international standards viz. IGES (Initial Graphics Exchange Specification) and STEP (ISO 10303 – Standard for Exchange of Product Data) are explained with examples.

To illustrate mathematical concepts presented, solved examples have been included in the relevant chapters of the book. In addition some review questions and practice problems have been included for self study.

I gratefully acknowledge the cooperation and support received during the course of writing this book. IIT Bombay provided me academic freedom and encouraging environment to carry out research and teaching activities in CAD/CAM all along. I sincerely thank all my Ph. D and M. Tech students who enriched the research environment by way of discussions and implementation of new ideas. The staff in my Computer Aided Manufacturing laboratory always provided me dedicated support. In particular I would like to thank Ms. Ashwini Patil for her help in meticulously preparing the manuscript of this book.

Finally I wish to express my special thanks to my wife Varsha and son Sushrut for their patience and constant support while writing this book.

S S Pande

Contents

Chapter 1

Introduction

Computer graphics is an interesting field which enables the creation of virtual world using digital images. It provides users software tools to model 3D object shapes and transform them dynamically to create animations. Computer graphics is widely used in a variety of applications such as scientific data visualization, Computer Aided Design and Manufacturing (CAD/CAM), satellite imagery, weather forecasting, medical imaging, computer animation, advertizing, television and filmmaking.

This chapter will present a brief introduction to Computer Graphics and Product Modeling with focus on CAD/CAM applications.

1.1 COMPUTER GRAPHICS

Computer graphics is a subfield of Computer Science which is primarily concerned with the *synthesis, manipulation* and *visualization* of digital images. In a broad sense, it focuses on the computational techniques for the representation and manipulation of geometric and visual content of an image. Computer graphics overlaps with fields such as Computer Vision and Image Processing on several mathematical and computational issues.

1.1.1 Historical Perspective

The single event that marked the evolution of Computer Graphics as an important, new field was the publication of a Ph.D thesis in 1962 by Ivan F. Sutherland, a student at MIT. The thesis titled *Sketchpad: A man–machine Graphical Communication System* conclusively demonstrated that computer graphics was a viable, useful and exciting field of research with wide ranging application potential.

Several researchers from academic and industrial laboratories contributed subsequently to the evolution and growth of computer graphics as a discipline during the 60s. Steve Russel, another student at MIT created the first video game *Spacewar* using DEC computer PDP-1 which became an instant success. By the mid 60's, major corporations like TRW, IBM, Lockheed – Georgia, General Electric, Sperry Rand took active interest in computer graphics in terms of development of hardware and software. Dave Evans and I. Sutherland started the computer science program at University of Utah in USA. By 1970, it became the hub of pioneering research which contributed several fundamental algorithms in computer graphics.

Realizing the potential of computer graphics, application oriented research was carried out during the 70's. Prominent among these developments was the emergence of Computer Aided Design

(CAD) and Computer Aided Manufacturing (CAM) technology for the aircraft and automobile manufacturing industries. Today computer graphics has become the core of several application areas from Digital art, biology and drug design to video games and Virtual Reality.

1.1.2 Picture Creation in Computer Graphics

In computer graphics, creation and manipulation of digital images, termed as *Pictures*, is of paramount importance. Quality of the picture is dictated by the end application. For example in Computer Aided Drafting, the picture is mainly a line drawing which consists of simple graphics entities like Line, Arc, Text etc. Applications like Solid Modeling, geographical imaging, games and computer animation need realistic rendering of images (pictures) with colors, shades and light reflections. Animation needs refreshing such realistic images at a specific time (frame rate) to create the perception of object motion.

Figure 1.1 shows the process of picture creation in computer graphics. Two approaches are broadly followed as under.

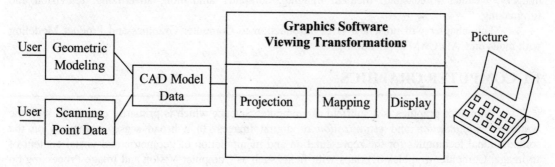

Figure 1.1 Picture Creation in Computer Graphics

Geometric Modeling

Geometric modeling provides utilities to the user to create 2D/3D object shapes in the user defined world. The modeler has a predefined set of geometric entities (primitives) like *Point, Line, Arc, Curve, Solids,* and a set of tools to *create, edit,* and *transform* them. Advanced modelers like Feature based systems enable user to model parts in terms of functionally relevant features for specific application domains like machined, die cast or sheet metal parts. Using the geometric modeler, user can create, manipulate, store and retrieve part and assembly models for CAD/CAM applications. Techniques of geometric and solid modeling are explained at length further in the book.

Image Creation from Points

Quite often it is required to create digital images (pictures) from data which has been obtained from satellite imaging, video, model (part) scanning or conduction of some scientific experiment. In industrial product development scenario, designers sometimes create an aesthetic clay model of the product such as an automobile. This activity is termed as *styling*. The clay model created by artist is digitally scanned and its shape is generated in the CAD modeling software from the points collected.

The point data or the CAD model so created is used further in CAD/CAM applications such as FE analysis, assembly interference checking, CNC programming or Rapid Prototyping. In medical imaging, human organs are scanned by X-rays, MRI or Laser probes and the point data is used for the construction of digital images (models) which are used for diagnosis.

The CAD model data generated either by the geometric modeler or from the scanned points is processed by the graphics software to create image on the display device.

1.2 PRODUCT DEVELOPMENT AND CAD/CAM

Due to global competition, manufacturing industries worldwide are facing several challenges such as shorter product life cycles, frequent design revisions, requirements of higher productivity and product quality and the need for shortest time to market. To meet these challenges, Computer Aided Design and Manufacturing (CAD/CAM) technology has been developed in the last three decades.

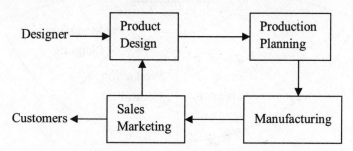

Figure 1.2 Stages in Product Development

Figure 1.2 shows the stages in a typical product development scenario. Activities can be conceptually grouped into four categories viz *Product Design, Production Planning, Manufacturing, Sales and Marketing*. These can be further classified into specific functional activities.

CAD/CAM technology essentially focuses upon automating and integrating these activities to create *flexibility* and *agility* in the product development cycle. Computer Aided Design (CAD) focuses on activities such as Conceptual Design, Detailed Design, Engineering Analysis, Assembly/ Interference analysis, Drafting while Computer Aided Manufacturing (CAM) deals with CNC programming, Robot planning, Factory management system involving Production Planning, Materials, sales, forecasting etc. CASA/ SME (Computer and Automation Systems Association division, Society of Manufacturing Engineers, USA) proposed the concept of Computer Integrated Manufacturing (CIM) which essentially aims at integrating the *Technical* and *Business* functions of an enterprise. Figure 1.3 shows conceptually various activities in CIM. Product Design, Analysis and Simulation form important activities among them. In the context of this book, Computer graphics and Product Modeling techniques for CAD/CAM applications are focused upon.

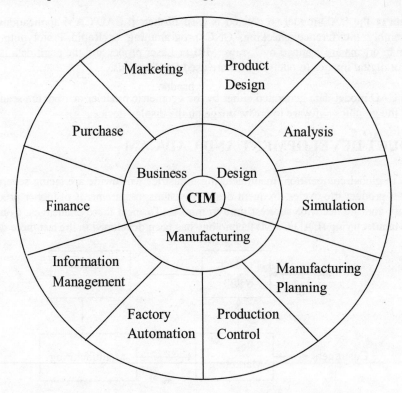

Figure 1.3 Activities in Computer Integrated Manufacturing (CIM)

1.2.1   Product Modeling

Product modeling is the heart of any CAD/CAM activity as it creates a central repository of product data which drives the downline application tasks. Product data can be broadly classified into two categories viz. *Geometric* and *Non Geometric*. Geometric data is concerned with product size, shape, features, constituent parts, their topology, surfaces, curves, vertex coordinates, to name a few. Non geometric data constitutes part name, number, production volume, material, part list (BOM), sales/ marketing data which is important for the business functions of the organization.

Efficient representation of product data is very important for the seamless integration and agility of the business functions of an organization. STEP is an evolving International standard for the efficient representation and sharing of Product Data in the electronic format. Due to stiff global competition, several forward looking industries worldwide, are implementing STEP standard to enable integration of product development and other business functions. This, in essence, is termed as *Product Life Cycle Management*.

1.3   PRODUCT LIFE CYCLE MANAGEMENT

Product Life Cycle Management (PLM) deals with the *creation*, *modification* and *exchange* of product information throughout the lifecycle of a product. It represents an all-encompassing vision for managing product data relating to the design, production, support and ultimate disposal of the manufactured goods.

During the past decade, manufacturers of industrial machinery, consumer electronics, instrumentation, packaged goods and other complex engineered products have realized the benefits of PLM and are thus, implementing it in increasing numbers.

1.3.1 Purpose

PLM can be thought of as both a repository for all information that affects a product as well as a communication process between product stakeholders viz engineering, manufacturing, marketing and field service. It is the first place where information from engineering and business functions encompassed by Computer Aided Design (CAD) , Computer Aided Manufacturing (CAM), Computer Aided Engineering (CAE), Enterprise Resource Planning (ERP) Customer Relationship Management (CRM), Manufacturing Execution System (MES), to name a few is integrated and shared between designers, manufacturers, suppliers, and customers.

1.3.2 Elements of PLM

PLM in essence, is a collection of digital solutions for various engineering and business applications. They can broadly be categorized as under

- Product and Process Structure creation
- Program management
- Lifecycle change and configuration Management
- Engineering Collaboration.

Product and Process structure enable the management of product and process data for all phases of the product life cycle. It encompasses process (project) related documentation, CAD design drawings, BOM, 3D model mockup data , etc to create a universal data model catering to the management, users and partners. Program management targets new product development and attempts to optimize processes from project planning, execution to progress analysis. It coordinates with accounting, sales, distribution, production planning, materials and maintenance activities. Lifecycle change and configuration management support all central logistic processes and provide mechanisms to enable and track product/assembly/configuration changes at a higher level. They govern product quality and economics of business process. Engineering collaboration enables integration of external and internal partners for an efficient product development scenario. It follows sharing of knowledge using Concurrent/ Collaborative principles.

1.3.3 Implementing PLM

PLM software developers offer range of solutions to industries to implement PLM. These are essentially based on the International standard ISO 10303 termed as STEP- Standard for Exchange of Product Data.

STEP provides variety of Application Protocols (AP) such as AP214 for automotive mechanical design, AP 212 for electromechanical design and installation, AP232 for technical data packaging core information and exchange, AP 203 for configuration controlled 3D design of parts and assembles to name a few. STEP file format and APs are discussed at length further. (Chapter 9)

1.4 SUMMARY

This chapter presented an overview of Computer Graphics and Product Modeling with a focus on product development using CAD/CAM. Subsequent chapters will present in details, the fundamentals of computer graphics, computer aided geometric design, geometric and solid modeling techniques and Product Data standards.

Chapter 2

Basics of Computer Graphics

Computer Graphics forms the core of any CAD/CAM activity. It enables a designer to interactively synthesize part shapes, carry out their functional analysis and simulate the performance of the assembly/ process. Computer Graphics in essence, facilitates the creation of this *Virtual World* for digital prototyping.

This chapter will introduce some key concepts and present an overview of the Computer Graphics systems.

2.1 ELEMENTS OF A GRAPHICS SYSTEM

In a computer graphics environment, *picture* is of paramount importance. Figure 2.1 shows the conceptual architecture of a graphics system indicating various modules in the picture generation process.

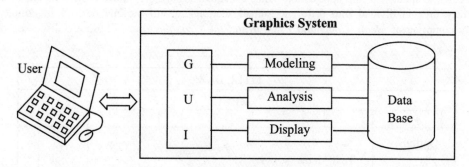

Figure 2.1 Architecture of a Graphics System

User interacts with the system with the help of familiar interactive devices like a *mouse* or the *keyboard*. The system presents to the user a graphical layout termed as Graphical User Interface (GUI) which is specific to the application task. User interacts with the system using the facilities provided by the GUI. The data inputted by the user is processed by the graphics application software and the picture of the desired model, process or graphical entity is finally displayed on the monitor to the user.

Figure 2.1 shows that there are four conceptual parts of the graphics software viz *GUI, Modeling, Analysis, and Display*. Database acts as a central repository which stores various data files. It provides integration and data sharing among the four elements of the graphics system during the user interaction. Majority of the algorithms are written in the graphics/ application software. However a few are a part of the graphics hardware or firmware to enhance the speed of picture processing and display.

A graphics system essentially comprises of two components viz. hardware and software. Hardware elements include *Output* devices such as display devices (monitors), printer, plotter and *Input* devices such as a keyboard, mouse, tablet, joystick or trackball. Detailed information on their construction and relative advantages/ limitation can be found in literature. Input /output devices commonly used in practice will be discussed here.

2.2 GRAPHICS SYSTEM HARDWARE

2.2.1 Input Devices

User interacts with the graphics system through the input devices. The most familiar input device is the keyboard. User types in commands using the keyboard which are displayed as *Text* on the screen and interpreted by the computer during operation. Though efficient for text input, the keyboard is a very abstract and unfriendly device to draw shapes in graphics environment. As a result for efficient graphics interaction many input devices such as mouse, trackball, digitizer (tablet) have been developed over the years.

In what follows, basic concepts of user- computer interaction (GUI) and the working of mouse will be presented.

2.2.1.1 Graphical User Interface (GUI)

During operation, user interacts with the computer through the Graphical User Interface (GUI) which is specific to the application being executed. Figure 2.2 shows typical layout of a GUI. The screen area is divided into different logical sections to suit various desired functions. They typically comprise of *Menu, Command Prompt and the Graphics area*. These are discussed one by one.

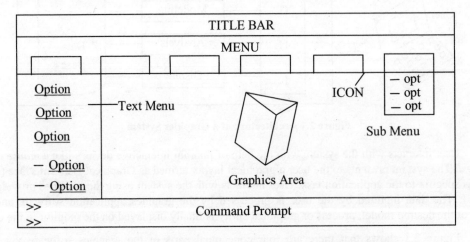

Figure 2.2 Typical Layout of a Graphical User Interface (GUI)

Menu

Menu is the list of options provided by the software to the user. Menu options are displayed to the user in the menu area in two styles.

- Text based menu items termed as *Hot Links*.
- Graphics (picture) buttons termed as *Icons*

GUI can have either or both of these styles based on the user interactivity and friendliness desired. If the options in the menu are small in number, all of them are displayed in the GUI. However this rarely is the case.

Generally the menu options are organized in the Hierarchic fashion and are displayed as functional groups. On clicking a specific icon, the menu options in that category scroll down for the user to pick from. Figure 2.2 shows a typical scroll down sub menu.

Command prompt

The command prompt area in GUI is generally kept at the bottom of the screen (Figure 2.2). Text commands entered by the user are displayed there and executed in the command mode by the system. Command prompt option is generally used by the experienced users of the CAD system who know which command is to be used in a particular design operation. They do not want to wade through the cascade of menus and submenus to reach to the desired command. Command based operation often enhances the user productivity.

Graphics area

It is the area on the screen where the picture (CAD model) is displayed. User can interact with the model using the input device. Figure 2.3 shows various activities carried out by the user.

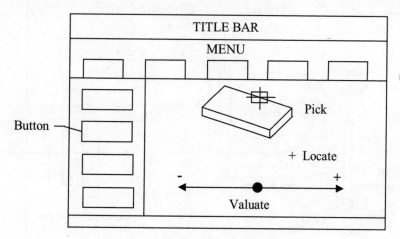

Figure 2.3 Logical Input Interactive Functions

The activities can be logically classified into four categories as under.

<u>Locator</u>	– To locate cursor at a point
<u>Valuator</u>	– To evaluate *Sign* and *Value*
<u>Button</u>	– To choose a menu item
<u>Pick</u>	– To select a graphics (picture) entity

All input devices such as a mouse, track ball, tablet etc implement the above four interactive actions. Mouse is the most widely used input device among the various input devices. Working of a mouse is described below.

2.2.1.2 The Mouse

Mouse is a user friendly input device connected to the computer. Mouse rests on a flat surface (pad, table) and can be moved in X-Y directions on the surface by the user. Movement of the mouse is transformed electronically into the movement of the cursor on the computer screen. User can thus, get an immediate visual feedback on the action of moving the mouse. Mouse generally has 2 or 3 buttons on the top surface which can be used to select items in the menu to communicate actions to the system.

Mouse was invented by Dr.Douglas Engelbart, a Professor with Stanford Research Institute, California, USA in 1964.The unit had one button on the top and two wheels at the bottom which tracked the horizontal and vertical movements. A cable transmitted signals from the device to the computer. Looking at the long cable, a member of Dr. Engelbart's team remarked that the device looks like a *mouse* and the name stuck even today. The development of personal computers stimulated the development of mouse. Various designs were developed by Xerox, Macintosh, Apple, Video game developers and even at NASA for flight devices. Figure 2.4 shows the inside details of a typical Opto-mechanical mouse commonly used in practice.

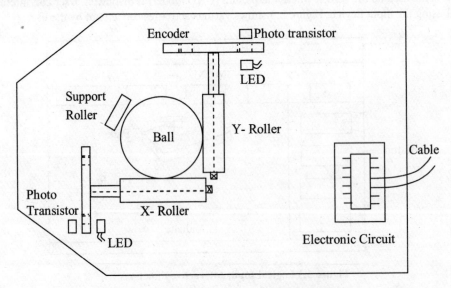

Figure 2.4 Construction of an Optomechanical mouse

The mouse has a hard plastic body which the user maneuvers across the pad / flat surface. A rubberized ball protrudes out of its bottom surface and remains in contact with the pad surface all along. As the mouse is moved, the ball turns in its socket inside the body. The ball is held in position by a support wheel and two cylindrical rollers perpendicular to each other (Figure 2.4)

Rotation of the ball causes rotation of the two rollers (shafts) which, in turn, track the horizontal (X) and vertical (Y) movement of the mouse. At the end of each roller (shaft), an encoder is mounted as shown. The encoder has a disc with slots (holes). A light emitting diode (LED) is fitted on one side of the wheel and a photo transistor (receiver) on other side. Infra red signals from the LED are received by the phototransistor. When the encoder disc turns along with the roller shafts, the signal from the LED is intercepted periodically depending upon the pitch of slots (holes) in the disc and its speed of rotation. The photo transistors translate these light and dark bands into electrical pulses which are processed by the on board electronics circuit and finally transmitted to the computer via the mouse cable. The pulses can tell the direction of movement Left/ Right/ Up/ Down and accordingly the computer instructs the cursor to be moved on the computer screen.

Each mouse design has its own software called as mouse driver. The mouse driver tells the computer how to interpret the signals received from the mouse in terms of the direction (X, Y), speed of movement and the commands clicked by the buttons on the mouse. Some mouse drivers allow users to customize the mouse in terms of assigning some functions to mouse buttons, adjusting the speed and resolution of the mouse etc.

Innovations are constantly being done on making mouse more and more user friendly and interactive. Notable among these include optical mouse without ball, wireless (tail free) mouse, Internet mouse with scroll wheel between buttons to name a few. Track ball is an inverted mouse wherein the user rolls the ball to move the cursor. These are used in Laptop computers and game stations.

2.2.2 Display Devices

Computer monitor is an important output device in the graphics system as the picture is finally displayed on it. Applications such as CAD/CAM, animation, entertainment have different requirements for the output picture in terms of resolution, color realism and dynamic picture update. Monitors, thus, need to have functional characteristics like high resolution, good picture quality, realism in image rendering (color, shading) and fast dynamic display.

Currently display monitors based on two different technologies are used. These are Cathode Ray Tube (CRT) and Liquid Crystal Display (LCD) technology. Desktop graphics systems have been using CRT monitors for the last 3 decades but the scenario is changing fast towards desktop LCD monitors. Laptop computers have LCD monitors since their inception a decade ago.

In what follows, a brief overview of CRT and LCD monitors for graphics system will be presented from the functional (user) considerations.

2.2.2.1 CRT Display Monitors

These monitors primarily use the Cathode Ray Tube (CRT) technology in which the screen is coated with high fluorescence phosphor. Beam of electrons strike the screen to generate a spot which is displaced by changing electric signals applied to the CRT tube. Due to the persistent phosphor, the image drawn by the electron spot appears to stay on the screen. CRT technology has been used since almost the inception of computer graphics in mid 60s.

CRT monitors fall into three categories as under

- Direct View Storage Tube (DVST) Graphics displays
- Calligraphic refresh Graphics displays
- Raster Refresh Graphics displays

DVST is conceptually the simplest and oldest of the CRT displays. The screen has high persistence phosphor causing the image to remain on screen for a very long time. However this is also a disadvantage as it is not possible to selectively erase a picture or part thereof. The entire screen needs to be flooded, erasing the whole picture in the process. The DVST display provides excellent picture quality, flicker free operation and accurate line drawing but the level of interactivity is poor.

In comparison with DVST, the Calligraphic refresh display has a low persistent phosphor screen. The image drawn by the electronic spot on the screen needs to be refreshed often (at least 30 times a second) to maintain the persistence of vision. The calligraphic displays are often called as Random Scan or Vector displays. They are basically line drawing displays.

For their operation, Calligraphic displays essentially need a *Display Buffer* and a *Display* controller in addition to the CRT (Figure 2.5). Picture processor routines implemented in hardware compute data on coordinates, transformations, projections etc and stores in the buffer all the information of the picture to be drawn. The display controller steps through the buffer to display vectors in the picture and at the end of the refresh cycle automatically reset to the top of the buffer to step through again. Needless to say that if the picture is too complex (many vectors) or the refresh rates fast, the display will cause flicker. Special techniques are needed in the design of the buffer to address such problems.

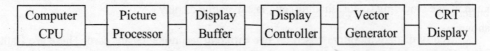

| Computer CPU | Picture Processor | Display Buffer | Display Controller | Vector Generator | CRT Display |

Figure 2.5 Conceptual modules for a calligraphic refresh display

Calligraphic displays provide capabilities of dynamic picture update i.e. the contents of the buffer can be updated on the fly during refresh. Both DVST and Calligraphic displays were monochromatic. With the advent of color displays, Raster graphics displays have became more popular in the last two decades. Today Raster graphics color displays are commonly used on all desktop graphics systems.

2.2.2.2 Raster Graphics CRT Displays

Both the DVST and the Calligraphic displays are line drawing devices which enable drawing line from one addressable point to other addressable one. In contrast, the Raster display is similar to an array or matrix. Each element of this matrix is a picture element termed as *Pixel*. Picture on the screen is composed of pixels which can be individually made bright with different color and intensities. In essence, the raster CRT device is a *Point Plotting* device against the line drawing one.

The screen will be composed of discrete pixels based on the resolution of the screen. Due to this characteristic, it is not possible to directly draw a straight line from one addressable point (pixel) to another. Figure 2.6 shows that only Horizontal, Vertical and lines at $45°$ can be accurately drawn (Figure 2.6 A). All other lines need to be approximated as a series of pixels to *reasonably* represent the shapes. Figure 2.6 B shows the possible candidate pixels to be chosen for approximating the line.

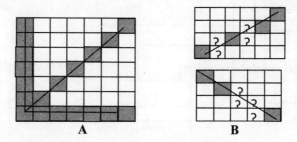

A   **B**

Figure 2.6 Raster Approximation of a line

The process of approximating the graphics entity (Line, Circle) by discrete pixels is termed as *Rastesrization*. Due to rasterzrion, all lines will appear to have a stair step effect. These are called as aliases or jaggies. Raterizing algorithms need to take care of the aliasing effect to create *reasonable* approximations of lines. Typical raterization algorithms will be discussed further in this chapter under Graphics software.

Frame Buffer

Frame buffer is the part of computer memory which drives the raster. Conceptually frame buffer can be considered as a matrix which has one memory bit for each pixel on the screen. Figure 2.7 shows a single plane frame buffer for a black and white CRT graphics device.

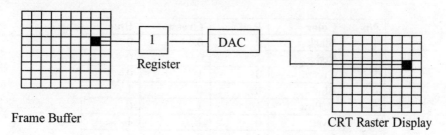

Frame Buffer   CRT Raster Display

Figure 2.7 Single bit plane frame buffer raster CRT display

Picture is built up in the frame buffer one bit at a time. The amount of memory needed is called as the bit plane. For example, if the raster has 512 x 512 pixels, it would need 2^{18} (512 x 512) bits of memory for a single plane black and white device. Since the frame buffer is digital while the pixel on the screen is an analog device, a digital to analog (DAC) converter is required (Figure 2.7)

Instead of a single plane, multiple bit plane frame buffers can be used. The addition of bit planes provides a very important facility to incorporate the color and/ or the grey levels to the pixel. Figure 2.8 shows the concept of a N bit plane grey level frame buffer for a black and white CRT monitor. The binary value of each corresponding pixel in the N bit frame buffer are fed to a register. The resulting N-bit binary word is interpreted as the intensity of the corresponding pixel on the raster screen. There will be thus, 2^N grey levels between 0 (dark) to 1(full white) for each pixel. Figure 2.8 show the arrangement for N=3 giving 8 (2^3) grey levels between the dark and bright.

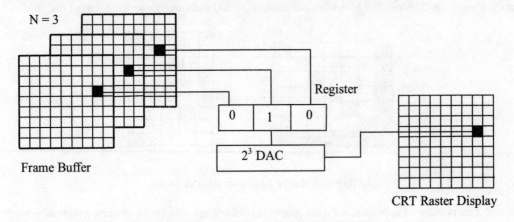

Figure 2.8 A 3 bit plane grey level frame buffer

A similar arrangement can be used for a color raster. There are three primary colors Red (R), Green (G), and Blue (B). Figure 2.9 shows the 3 bit plane color frame buffer arrangement. Each bit plane drives one color gun R, G, or B. The pixel on screen will be controlled by the 3 bit word based on the combination of R, G and B colors. Table 2.1 shows the color combinations and the resulting colors for a 3 bit plane color frame buffer

Table 2.1
Color Combinations in 3 bit plane frame buffer

No	Color	Red	Green	Blue
1	Black	0	0	0
2	Red	1	0	0
3	Green	0	1	0
4	Blue	0	0	1
5	Yellow	1	1	0
6	Cyan	0	1	1
7	Magenta	1	0	1
8	White	1	1	1

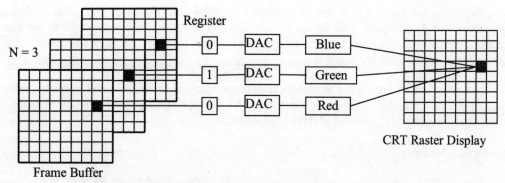

Figure 2.9 Simple color frame buffer raster CRT display

Figure 2.8 and 2.9 show how a 3 bit plane frame buffer can effectively change the intensity (grey level) and color of the pixel. The two concepts are integrated to change both color and intensity in a color raster device. Figure 2.10 shows a 24 bit color frame buffer. For each primary color R, G, B, 8 bit planes are implemented. Thus each color has 2^8 intensity levels. These 8 bit binary words drive each color gun through individual 8 bit DAC. Further the three primary color intensities are combined to yield $(2^8)^3 = 2^{24} = 16,777,216$ possible colors

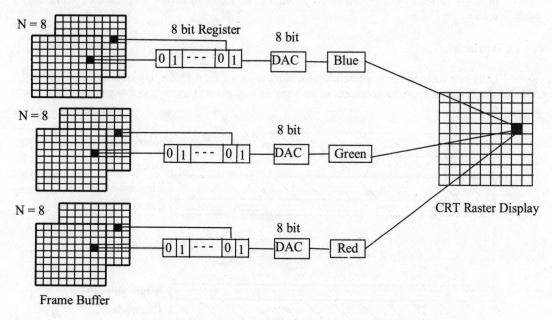

Figure 2.10 A 24 bit plane color frame buffer raster CRT display

Thus a 24 bit plane frame buffer is capable of producing 16 million colors. This is the full color frame buffer.

The full color frame buffer can be expanded by using color lookup tables. This facilitates customization wherein groups of colors from the possible palette can be chosen at a time by the user. To achieve dynamic picture update in real time, pixels in groups of 16, 32, 64 etc are accessed i.e. all bit planes for a pixel are accessed together.

The raster color frame buffer provides capability to manipulate the color and intensity of each individual pixel of the image. As a result the raster CRT displays are capable of producing real life dynamic pictures required in computer graphics and animation.

2.2.2.3 LCD Monitors

Though CRT technology is quite mature and widely in use for the past 30 years, it has some basic limitations. The CRT monitors are quite bulky and consume lot of power. They are suitable for desktop applications only.

Liquid crystal display (LCD) technology was developed in the last 15 years as an alternative to the CRT. In the initial days LCD technology was limited to monochromatic displays which are widely used in calculators and digital watches. However in the last few years LCD color monitors have been developed which are used in laptop computers and home TVs.

In what follows, the principle of operation of LCD monitors is presented from user's perspective

The LCD principle

LCD primarily uses the principle of *Polarization* of light to display objects. Like the raster CRT, the LCD screen (monitor) consists of an array of tiny pixels (segments) that can be manipulated to effect the display.

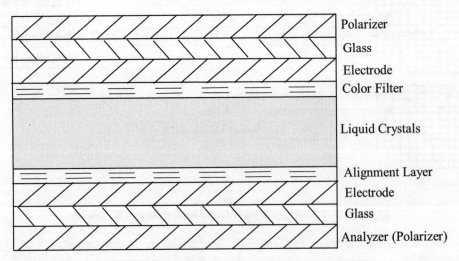

Figure 2.11 Construction of a Liquid Crystal Display

Figure 2.11 shows the construction of a LCD screen. It has a *Polarizer* and an *Analyzer* on the top and bottom with liquid crystal molecules in between. The assembly has electrodes for supplying voltage during operation. The Polarizer and Analyzer are usually perpendicular to each other. The liquid crystal molecules have a property that when they come in contact with grooved surfaces having fixed orientations, they align parallel to the grooves.

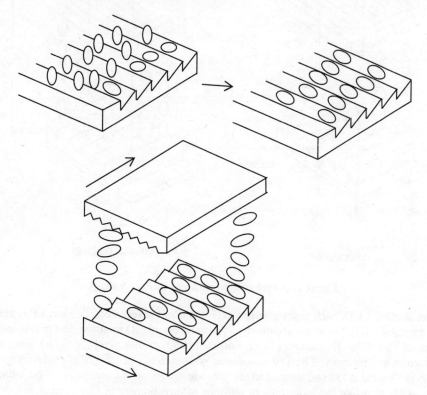

Figure 2.12 Alignment of liquid crystal molecules

Figure 2.12 shows the process of alignment of the liquid crystal molecules. It shows the structural arrangement of molecules when no voltage is applied. It can be seen that the molecules provide a *Twisted* structural arrangement from top to bottom. Figure 2.13 shows the path of light through the LCD under no Voltage condition. Polarized light enters the LCD from top and passes through the spacing of the molecular arrangement. In turn, the light gets twisted by $90°$. It finally passes out of the Analyzer at bottom which in effect, is another polarizer. Since the light passes through the display, it looks transparent. When the voltage is applied the molecules straighten out of their helix pattern and get aligned along the direction of voltage as shown in Figure 2.13. The polarized light now does not get twisted and so cannot pass through the analyzer. The LCD segment thus, looks opaque.

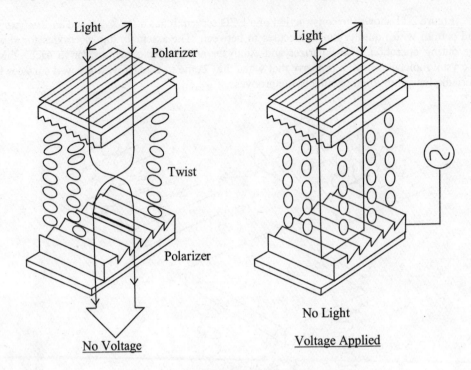

Figure 2.13 Path of light with and without Voltage

For a color LCD, each pixel consists of 3 color sub pixels Red(R), Green (G) and Blue (B) similar to the color CRT. These are arranged very closely on a LCD monitor. The pixels are controlled by a matrix of Thin Film Transistors (TFT). For each sub part of a pixel (R, G, B) there is a TFT to control the color and intensity. The TFT associated with each sub pixel controls the voltage applied to the equalizers (Figure 2.13) and thus changes the orientation of the molecules. The voltage control makes it possible to orient the molecules in millions of directions with fine control and thus produce corresponding intensities in the respective color tones. As a single pixel consists of 3 primary colors, the overall visual effect is the hundreds of thousands of different colors.

Advantages and limitations

LCD monitors offer many advantages compared to a CRT display. Important among them are as under

- Very light weight
- Sleekness and portability
- Very low power consumption
- No electromagnetic fields
- Longer life

LCD monitor, however, have a poor viewing range ($\pm 45°$). High resolution active matrix displays are expensive. However with advances in technology, these will soon be overcome. LCD monitor are the future of the graphics displays.

2.3 GRAPHICS SOFTWARE

As discussed earlier, user interacts with the graphics system through the GUI and carries out the modeling operations to create picture on the monitor (Figure 2.1). Graphics software enables the user to create GUI and write application software using some graphics kernel (core) commands.

Conceptually graphics software can be considered to have three functional components. The first one focuses upon the process of picture generation from modeling world to display. These are termed as *Viewing Pipeline Transformations*. The second component deals with the algorithms for display of pictures on monitors. These display algorithms are often implemented in graphics hardware or firmware to accelerate the display performance. The third component is, in essence, the graphic standards which provide the programmer utilities to write application software and enable exchange of graphics model data to other software for integration.

In what follows, mathematical basis of the viewing pipeline and raster scan graphics algorithms are presented. Product data standard is discussed at length in Chapter 9.

2.3.1 Viewing Pipeline Transformations

The objective of Viewing Pipeline algorithms is to primarily transform object models in 3D world into pictures on the graphics system device (monitor). Figure 2.14 shows the conceptual representation of the viewing pipeline transformations. It primarily comprises of three modules as under

- Projection Transformations
- Windowing and Clipping
- Window to viewport mapping

Various modules of the viewing pipeline are discussed one by one.

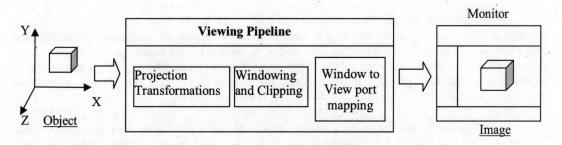

Figure 2.14 Viewing Pipeline

2.3.1.1 Modeling World

In computer graphics, user models the objects (scene) in the computational / mathematical world. The modeling space is a bounded space in R^3 defined by the user. In computer graphics and CAD/ CAM, *Right Handed* Cartesian Co-ordinate system shown in Figure 2.15 is used. It is termed as the World Co-ordinate System – **WCS.**

User defines the system of units to be followed (such as mm, meter, feet, inches, miles, microns) depending upon the application world to be modeled. The object models are normally created in the first quadrant/ octant in WCS where all the coordinate dimensions are positive. Suitable geometric transformations such as Rotation, Translation and Scaling etc can be done to position the object in 3D world. These are discussed at length in Chapter 3.

2.3.1.2 Projection Transformation

This is the first and most important viewing transformation. Projection operation aims at transforming the 3D object model in world to its 2D image on the viewing plane. Conceptually the projection operation is similar to taking the snapshot of an object using a camera. In computer graphics, a *Virtual Camera* is used to effect this operation

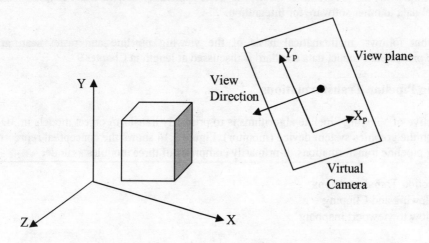

Figure 2.15 World Coordinate System (WCS) with Virtual Camera

Figure 2.15 shows the *Virtual Camera* located in the world. User defines the location and orientation of the camera in the WCS. The camera *looks* at the objects and based on the type of projection desired (Parallel / Perspective), it creates projection of the object on the viewing plane (Figure 2.15). The projected image of the object created on the view plane is 2D in nature and is represented on the view plane in its local coordinate system ($X_p - Y_p$). The 3D object model is represented in world (WCS) in terms of its Faces, Edges and Vertices etc.

Fundamentally, the object faces, edges, vertices and their interconnections (termed as object topology) do not change during the process of projection. The object vertex is represented in WCS as a position vector **P** with its coordinates x, y, z i.e $P(x, y, z)$. During projection, it gets transformed to $P_p(x_p, y_p)$ on the view plane in $X_p - Y_p$ coordinate system. All vertices of the object are transformed accordingly during the projection. Connecting them as per the object topology creates the projection of the object on the view plane.

Mathematical basis of projection transformations is presented in details in Chapter 4.Techniques of geometric and solid modeling of 3D objects are discussed at length in Chapters 7 and 8.

2.3.1.3 Windowing and Clipping

Window is a user defined rectangular area defined on the view plane ($X_p - Y_p$) which contains the final image to be projected on the display device. Figure 2.16 shows the projected image and the window defined by its diagonal corners (x_{w1} y_{w1}) and (x_{w2} y_{w2}). If the size of the window is smaller than the extents of the picture, (Figure 2.16) portion of the picture lying outside the window limits will be cropped and will not be seen on display. This operation of picture cropping is termed as *Clipping*.

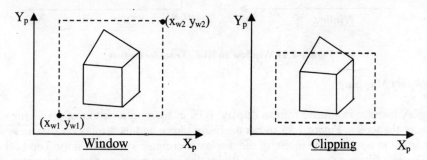

Figure 2.16 Windowing and Clipping Transformation

By adjusting the size of the window with reference to the projected image, the size of the final picture on display can be changed. During operation, the user dynamically drags the corners of the window to change the size of image on display. This operation is termed as *Zooming*.

2.3.1.4 Window to Viewport Mapping

This transformation essentially aims at mapping the contents of the window on the projection plane onto a specific part termed as Viewport on the display device (monitor). This operation is carried out in two steps as under.
- Window to Normalized Device Coordinate (NDC) space
- NDC space to Viewport

These are explained one by one.

Window to NDC space

During the projection transformations, vertices of the 3D object are mapped on the 2D view plane (X_p-Y_p) plane. Figure 2.17 shows a typical projected point P_p (x_p, y_p) on the view plane. It lies in the window defined by corners (x_{w1} y_{w1}) and (x_{w2} y_{w2}). The point P_p (x_p, y_p) is to be mapped to the Normalized Device Coordinate (NDC) frame defined by X_N - Y_N. The mapped point P_n (x_n, y_n) is given by

$$x_n = \frac{x_p - x_{w1}}{x_{w2} - x_{w1}}$$

$$y_n = \frac{y_p - y_{w1}}{y_{w2} - y_{w1}}$$

2.1

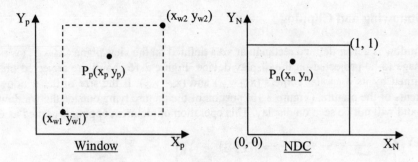

Figure 2.17 Window to NDC Transformation

NDC to Viewport Mapping

Display monitor is a raster refresh display. It is, in essence, a matrix of pixels (m x n) as per the resolution of the screen. Figure 2.18 shows the raster screen and its display coordinate system $X_D - Y_D$. It is important to note that the origin of the display coordinate system is at the Top Left corner of the screen. The coordinate system is integer in nature due to the pixel matrix on the screen. Since the resolution of the monitor may be variable as per the specific one used, the viewing pipeline maps the window contents to NDC for general transformations. NDC to viewport mapping is specific to the display.

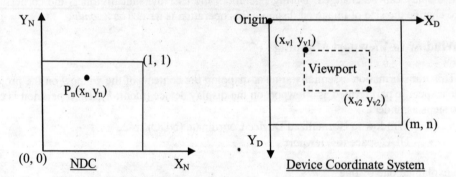

Figure 2.18 NDC to Device Coordinate System Transformation

Viewport is a specific rectangular area on the device wherein the contents of window are finally mapped. It is specified by diagonal corners (x_{v1}, y_{v1}) and (x_{v2}, y_{v2}). As an illustration, considering that NDC space is mapped to the entire screen (device), the transformed point $P_d (x_d, y_d)$ on the display (Figure 2.18) will be given by

$$x_d = x_n.m$$
$$y_d = (1 - y_n).n \qquad\qquad\qquad \textbf{2.2}$$

where (m, n) is the monitor resolution in $X_D - Y_D$.

The viewing pipeline thus, transforms a point P(x, y, z) in 3D object world (WCS) to the point $P_d (x_d, y_d)$ in the 2D image world on the device.

2.3.2 Raster Scan Graphics Algorithms

As discussed earlier, the raster display is similar to a matrix of pixels which can be made dark/ bright. The picture on the graphics display will thus, be approximated to the resolution of the screen. Algorithms for drawing graphics primitives like Line, Circle etc are usually implemented in hardware to improve the speed of display.

In what follows raster scan algorithms for Line drawing will be presented.

2.3.2.1 Line Drawing Algorithms

Rasterization is the process of determining which pixels should be made bright to provide best approximation for drawing a desired line. Figure 2.6 shows various situations of rasterization. For *Horizontal, Vertical* or *45°* lines, the choice of raster elements (Pixels) is obvious. However for other line orientations (Figure 2.6 B) it needs some decision.

The general requirements for line drawing are as under

* Straight lines must appear as *Straight* lines.
* They must start and end accurately.
* Lines should have constant brightness along their length.
* Speed of drawing lines should be fast.

It is very difficult to satisfy all the above requirements simultaneously. As a result, compromises are sought in drawing lines which give reasonable accuracy, brightness and speed in drawing. Figure 2.19 shows two alternate ways of drawing a line. The candidate pixels are shown by ?. Both options however, provide uneven brightness. Special algorithms are needed to handle such problems.

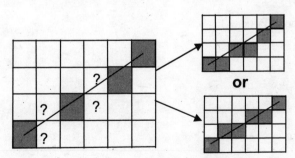

Figure 2.19 Rasterization of a straight line

Two most widely used algorithms for raster line drawing are Digital Differential Analyzer (DDA) algorithm and Bresenham's algorithm. These are discussed below.

Digital Differential Analyzer (DDA) Algorithm

The DDA algorithm operates on the principle of solving the differential equation of the line using incremental method.

Differential equation of a straight line is given by

$$\frac{dy}{dx} = \text{constant} \qquad \qquad 2.3$$

The equation can be written in the difference form as

$$\frac{dy}{dx} = \frac{\Delta y}{\Delta x} = m \qquad \qquad 2.4$$

where m is the slope of the line .

Let $P_1(x_1, y_1)$ and $P_2(x_2\ y_2)$ represent the end points of the line to be drawn on the raster screen. Needless to say that x_1, y_1, x_2, y_2 will be integers representing the pixel addresses (coordinates) Slope of the line is given by

$$m = \frac{y_2 - y_1}{x_2 - x_1} \qquad \qquad 2.5$$

Knowing a point $P_i(x_i\ y_i)$, successive point $P_{i+1}(x_{i+1}\ y_{i+1})$ can be computed by solving the equation 2.3. The recursion relation is given by

$$x_{i+1} = x_i + \Delta x$$
$$y_{i+1} = y_i + \Delta y \qquad \qquad 2.6$$

This relation is the basis of the digital differential analyzer (DDA). The choice of input variable to Equation 2.6 (Δx or Δy) is chosen based on the slope of the line m. For $m \in [-1, 1]$, Δx is chosen as 1 pixel unit. The steps in the DDA algorithm are enumerated below

Algorithm
- Choose initial point P_1 ($x_1\ y_1$)
- Set $X = x_1$
        $Y_{True} = y_1$
- Loop while ($X \leq x_2$)
- $Y = \text{integer } (Y_{True})$
- Plot pixel (X, Y)
- $X = X+1$
- $Y_{True} = Y_{True} + m$
- End loop

Integer is a function to convert Y_{True} real values to the integer data. Sometimes Round is also used to round off the value of Y_{True} to the nearest integer. The DDA algorithm needs to compute the slope (m) only once at the start of the computation. The DDA algorithm is simple and elegant but has two limitations as under
- The accuracy of line (pixel locations) is dependent upon the orientation (slope) of the line being drawn.
- The algorithm runs slowly as it performs real arithmetic using the floating point operations.

Bresenham's algorithm discussed in the next section is a more suitable raster line drawing algorithm.

Bresenham's Algorithm

Unlike the DDA, Bresenham's algorithm uses integer arithmetic only and hence performs significantly faster. The algorithm primarily computes optimum raster locations to represent a straight line. During computation, the algorithm increments by one unit either in X or Y direction depending upon the slope of the line. The increment in other direction is computed by the algorithm.

The key idea behind the algorithm is shown in Figure 2.20. Suppose the current location of the pixel is P(0,0) and the increment is in X direction by 1 unit. Line L passing through P (0, 0) is to be represented. At $\Delta x = 1$, two possible choices exist to locate the next pixel viz P (1, 0) or P (1, 1).

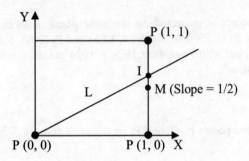

Figure 2.20 Concept of Bresenham's Line drawing algorithm

It is decided by examining the location of the point of intersection (I) of the desired line (L) with the line through P (1, 0) and P (1, 1). Midpoint M is chosen which represents the slope m = ½. There cases arise as to how I is located with respect to M.
Case I : I above M. (line slope > 1/2)
Case II : I coincides with M. (line slope = 1/2)
Case III : I below M. (line slope < 1/2)

For cases I and II, better pixel location is P (1, 1) while for case III, P (1, 0) is a better pixel location. To evaluate various cases, it is merely sufficient to check the sign of the relative location between I and M. This is termed as the *error* e which is used to decide the desired pixel locations. Various steps of the algorithm are as under

Algorithm
- Choose initial point P_1 (x_1 y_1)
- Set   X= x_1,
             Y= y_1
             e = -0.5, where e is the error term
- Loop while (x ≤ x_2)
- Plot Pixel (X, Y)
- Compute
             X = x + 1
             e = e + m

- If $e \geq 0$
  $$Y = y + 1$$
  $$e = e - 1 \quad \text{(re initialize error)}$$
  else $\quad Y = y_i$
- End loop

In essence, the error term e is the measure of Y intercept of the desired line (L) to be drawn at each raster element with reference to point M (Slope 1/2). Since only the sign of error term e is checked the algorithm runs faster. Illustrative examples are included in this chapter on DDA and Bresenham's line drawing algorithm.

2.4 EXAMPLES

1. Triangular facet of an object is projected on the view plane. Coordinates of the vertices in the projected plane (X_p - Y_p) are P_1 (1.75, 3.0), P_2 (6.18, 1.25) and P_3 (2.80, 5.15). Compute projections of these points on a raster display with resolution 1024 x 1024 pixels. Choose the size of the window appropriately. Draw a neat sketch.

Solution:

Figure 2.21 shows the points P_1, P_2 and P_3 on the view plane X_p - Y_p. Extents of the facet are

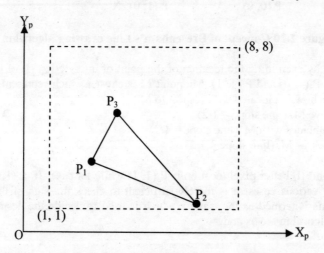

Figure 2.21 Facet and Window on view plane

$$X_{min} = 1.75 \qquad Y_{min} = 1.25$$
$$X_{max} = 6.18 \qquad Y_{max} = 5.15$$

The window should be chosen to be larger than the extents to contain the entire facet with proper aspect (length to width) ratio. Choosing the lower and upper diagonal corners of the window as (Figure 2.21)

$$x_{w1} = 1 \qquad y_{w1} = 1$$
$$x_{w2} = 8 \qquad y_{w2} = 8$$

Points on the view plane can be mapped to display device in two stages viz window to NDC (Equation 2.1) and NDC to view port (Equation 2.2). Assuming that the contents of window are to be mapped on the entire raster device, Equation 2.1 and Equation 2.2 can be combined. Point P_d (x_d, y_d) on raster device is given by

$$x_d = \frac{(x_p - x_{w1})}{(x_{w2} - x_{w1})} X_D$$

$$y_d = \left(1 - \frac{(y_p - y_{w1})}{(y_{w2} - y_{w1})}\right) Y_D \qquad\qquad \textbf{2.7}$$

For the chosen raster display, $X_D = Y_D = 1024$

Using Equation 2.7 the projected points on the raster device are computed. Table 2.2 lists the original and projected facet vertex coordinates. Figure 2.22 shows the projection of the facet on the raster display device.

Table 2.2
Projected vertices

Points	View Plane		Raster Device	
	x_p	y_p	x_d	y_d
P_1	1.75	3.00	109	731
P_2	6.18	1.25	758	988
P_3	2.80	5.15	263	417

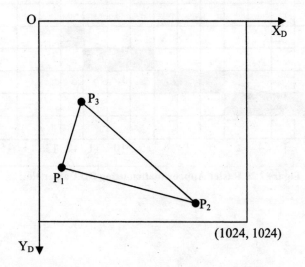

Figure 2.22 Facet projection on raster display

2. A line is to be drawn on a raster device from P_1 (2,2) to P_2 (12,5). Using the DDA algorithm compute the raster approximation to the line. Draw a neat sketch.

Solution:

The line is to be drawn from P_1 (x_1, y_1) to P_2 (x_2, y_2). Thus

$$x_1 = 2 \qquad y_1 = 2$$
$$x_2 = 12 \qquad y_2 = 5$$
$$\Delta x = x_2 - x_1$$
$$= 10 \text{ raster units}$$
$$\Delta y = y_2 - y_1$$
$$= 3 \text{ raster units}$$

Slope of the line m = $\dfrac{\Delta y}{\Delta x}$

$$m = 0.3$$

Since $\Delta x > \Delta y$, choose the increments of 1 raster unit in X direction.

Using the DDA algorithm, computations are carried out. Table 2.3 shows the results of computation. Figure 2.23 shows the exact line and its raster approximation using DDA.

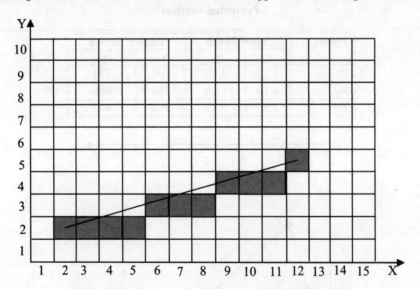

Figure 2.23 Raster Approximation using DDA algorithm

Table 2.3
DDA Algorithm

Iteration	x	y_{True}	y
1	2	2.00	2
2	3	2.30	2
3	4	2.60	2
4	5	2.90	2
5	6	3.20	3
6	7	3.50	3
7	8	3.80	3
8	9	4.10	4
9	10	4.40	4
10	11	4.70	4
11	12	5.00	5

3. A line is to be drawn on a raster device from P_1 (2,2) to P_2 (12,5). Using the Bresenham's algorithm, compute the raster approximation to the line. Draw a neat sketch. Compare the results with those from DDA algorithm (problem 2).

Solution:

As before, setting the initial values

$$x_1 = 2 \qquad y_1 = 2 , \ x_2 = 12 \qquad y_2 = 5$$
$$\text{slope } m = 0.3, \ \text{initial error e} = -0.5$$

Choosing increments of 1 raster unit in X direction, computations are carried out using Bresenham's algorithm. Table 2.4 shows the results of computation. Figure 2.24 shows the exact line and its raster approximation using Bresenham's algorithm.

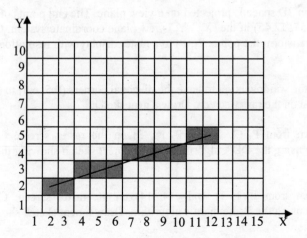

Figure 2.24 Raster Approximation using Bresenham's algorithm

Table 2.4
Bresenham's algorithm

Iteration	X	error (e)	$Y_{increment}$	Y
1	2	-0.5	0	2
2	3	-0.2	0	2
3	4	+0.1 (error reset to -0.9)	1	3
4	5	-0.6	0	3
5	6	-0.3	0	3
6	7	0 (error reset to -1)	1	4
7	8	-0.7	0	4
8	9	-0.4	0	4
9	10	-0.1	0	4
10	11	+0.2 (error reset to -0.8)	1	5
11	12	-0.5	0	5

Comparing Figure 2.23 and Figure 2.24 it is seen that the raster approximation for the same line given by the two algorithms are different. Bresenham's algorithm provides a better raster approximation to the line

2.5 REVIEW QUESTIONS

1. Explain why
   - Mouse is not an accurate input device for *Valuation*
   - Bresenham's line drawing algorithm is faster in operation than the DDA algorithm.

2. The edge of an object in 3D space is projected on a view plane. The end points of the line (edge) are $P_1(3.55, 1.28)$ and $P_2(10.57, 12.45)$ in the $(X_p - Y_p)$ view plane coordinate system. Choose appropriate size of the window and transform the points P_1, P_2 to a raster display with resolution 1024 x 768 pixels. Draw a neat sketch.

3. In the above problem, the window is set to have the diagonal corners (0.5, 6.5) to (10, 10). Compute the part of the picture seen on the raster screen. Draw a neat sketch.

4. A line is to be drawn from P_1 (2, 12) to P_2 (8, 3) on the raster screen. Compute the raster approximation to the line using the DDA algorithm. Draw a neat sketch showing the exact line and its approximation.

5. A line is to be drawn from $P_1(1, 1)$ to P_2 (4, 14) on the raster screen. Compute the raster approximation to the line using the Bresenham's algorithm. Draw a neat sketch showing the exact line and its approximation.

Chapter 3

Geometric Transformations

Geometric transformations play an important role in Computer Graphics and CAD/CAM as they enable a designer to position and orient objects during modeling. In particular, geometric transformations are used for a variety of applications such as computer aided drafting, 3 D part / assembly modeling, kinematic analysis, CNC programming, rapid prototype process planning.

In this chapter, mathematical basis for carrying out different types of geometric transformations will be discussed.

3.1 TYPES OF GEOMETRIC TRANSFORMATIONS

In a CAD system, 3 D object is represented In terms of its geometry and topology. For example a Tetrahedron shown in Figure 3.1 is stored in terms of its 4 faces, 6 edges and 4 vertices connected in a particular manner. The relation between Faces, Edges and Vertices is termed as the Object Topology. For a chosen object shape, the topology remains same irrespective of the size of the object. Coordinates of the Vertices govern the size of the object;

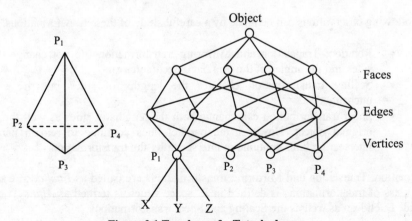

Figure 3.1 Topology of a Tetrahedron

Topology of the object does not change during Geometric Transformations, but the coordinates of the vertices get transformed. Thus, during the object transformation, the fundamental geometric entity undergoing transformation is the vertex. It is represented as a Position Vector **P** [x,y,z] where **P C R**^{3}. Geometric transformation is, in essence, finding a function **f : P -> P'** where **P'** denotes the transformed position vector.

Figure 3.2 shows a unit square and its resulting shape under various 2 D transformations.

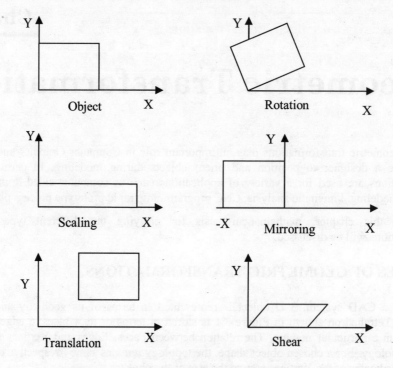

Figure 3.2 Types of Geometric Transformations

Following observations can be made by a careful study of these transformations.

- Rotation, Translation and Mirroring transformations do not change the lengths of sides and the angles of the original object (square).
- Scaling transformation changes the lengths of sides but the angles remain unchanged.
- Shear transformation can change both the lengths of sides as well the angles of the original object. However the parallel lines continue to remain parallel and the intersecting lines continue to intersect after the transformation.

Rotation, Translation and Mirroring transformations are called as *Euclidean* transformations. A general class of transformations is defined in R³ space which is termed as *Affine* Transformations. They include Euclidean as well as the Scaling and Shear transformations.

In this chapter, mathematical basis of Affine transformations will be presented.

3.2 GEOMETRIC TRANSFORMATIONS IN 2 D

3.2.1 Cartesian Coordinate Transformations

Two techniques exist to carry out the geometric transformations.

- Stationary Coordinate Frame (Axes) , transform the Point
- Stationary Point, Transform the Coordinate Frame (Axes)

In Computer Graphics, the first technique is generally followed and will be discussed.

Rotation

Figure 3.3 shows the Rotation of a point **P [x,y]** about the Origin in the counterclockwise direction by an angle **θ.** The point **P** is at a distance of R from origin and makes an angle **α** with the X axis.

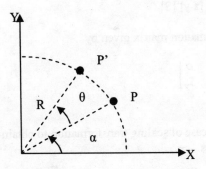

Figure 3.3 Rotation in 2D

Writing coordinate transformation equations,

$$X = R \cos (\alpha)$$
$$Y = R \sin (\alpha)$$

3.1

$$X' = R \cos (\alpha + \theta)$$
$$Y' = R \sin (\alpha + \theta)$$

3.2

Substituting Equation. 3.1 in 3.2 and simplifying,

$$X' = x \cos (\theta) - y \sin (\theta)$$
$$Y' = x \sin (\theta) + y \cos (\theta)$$

Putting in matrix form, the transformation equation is

$$P' = P [R]$$

$$[x' \ y'] = [x \ y] \ [R]$$

Where R is the Rotation transformation matrix given by

$$R = \begin{bmatrix} Cos \ \theta & Sin \ \theta \\ -Sin \ \theta & Cos \ \theta \end{bmatrix}$$

3.3

Scaling

A point **P [x,y]** can be scaled about the *Origin* using the scaling factors S_x, S_y. They need not be equal. Writing coordinate transformation equations,

$$X' = x . S_x$$
$$Y' = y. S_y$$

Putting in matrix form, the transformation equation is

$$P' = P \ [S]$$
$$[x' \ y'] = [x \ y] \ [S]$$

Where S is the Scaling transformation matrix given by

$$S = \begin{bmatrix} S_x & o \\ 0 & S_y \end{bmatrix}$$

3.4

Mirroring

Mirroring is a special case of scaling transformation to obtain the image (reflection) of a point **P** about X or Y coordinates axes.

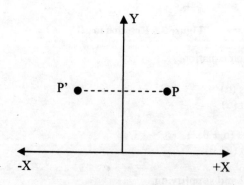

Figure 3.4 Mirroring Transformation

P' is the mirror image of a point **P** about **Y axis** (Figure. 3.4). Writing the transformation equations

$$P' = P \ [M]$$
$$[x' \ y'] = [x \ y] \ [M]$$

Where **M** is the mirroring transformation matrix given by

$$M = \begin{bmatrix} -1 & 0 \\ 0 & 1 \end{bmatrix}$$

3.5

Translation

Figure 3.5 shows the translation of a point **P** to **P'**

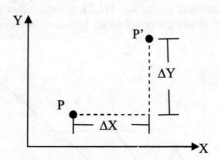

Figure 3.5 Translation in 2D

Writing the coordinate transformation equations

$$X' = x + \Delta x$$
$$Y' = y + \Delta y$$

3.6

where $\Delta x, \Delta y$ represent the translation distances in X, Y respectively.

Generalized Transformation Matrix

It is preferable to combine all the above transformations into a single transformation matrix so that matrix operations can be effectively used during computations.

$$P' = P [A]$$

Where **A** is a generalized transformation matrix

$$A = \begin{bmatrix} a11 & a12 \\ a21 & a22 \end{bmatrix}$$

It can be seen that Rotation, Scaling and Mirroring transformation matrices can be represented as **[A]** but the Translation equations (**Equation 3.6**) cannot be represented. The Cartesian coordinate system is thus, not able to represent a generalized unified transformation matrix.

This difficulty is solved by using Homogeneous Coordinate System.

3.2.2 Homogeneous Coordinate System

In Homogeneous coordinate system (HCS), a point in Cartesian coordinate system is mapped to a higher dimensional space by using a scalar w. Thus

$$P[x \; y]:\rightarrow P^h[\;wx \; wy \; w\;]$$
$$P[x \; y \; z]:\rightarrow P^h[wx \; wy \; wz \; w], \; w \text{ is a scalar, } w \neq 0$$

A point in 2D Cartesian space say **P(1,2)**, is equivalent to $P^h(\;1,2,1), P^h(\;5,10,5), P^h(1000, 2000, 1000)$ etc. Figure 3.6 shows the relationship between Cartesian and Homogeneous coordinates in a conceptual manner.

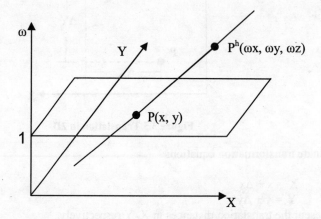

Figure 3.6 Homogeneous and Cartesian Coordinate System

It is seen that mapping from $\mathbf{P}:\rightarrow P^h$ is one to many and is thus, non unique. Mapping from $P^h:\rightarrow \mathbf{P}$ is obtained by

$$x = wx/w \;, \; y = wy/w \;, \; z = wz/w, \; w \neq 0$$

Homogeneous coordinate system offers many advantages in computation in terms of avoiding under/overflows errors, unifying operations/functions etc. They are widely used in computer aided geometric design.

In this book, Homogeneous coordinate system (HCS) will be used throughout to represent and transform curves, surfaces and 3 D object models.

3.2.3 Composite Transformations

Using Homogeneous coordinate system, point **P[x, y]** will be represented as $P^h[wx \; wy \; wz \; w]$. Without loss of generality, w can be chosen as 1. Dropping the superscript h, the point is represented as position vector **P[x, y, 1]**.

The generalized 2D transformation equation is

$$P' = P\,[A]$$
$$[x'\ y'\ 1] = [x\ y\ 1]\,[A]$$

Where **A** is the generalized affine transformation matrix

$$A = \begin{bmatrix} a11 & a12 & 0 \\ a21 & a22 & 0 \\ a31 & a32 & 1 \end{bmatrix}$$
3.7

It can be seen that all the 2D geometric transformation operations discussed so far like Rotation, Scaling, Mirroring and Translation (Equation 3.3-3.6) can be represented by a single matrix **A** (Equation 3.7).

Composite transformation connotes a situation in where an object (point) undergoes a series of transformations in succession. These can be easily handled by concatenating the variants of **[A]** corresponding to the transformations desired.

$$P' = P.\,[A_1]\,[A_2]\,[A_3]\,\ldots\ldots\ldots\,[A_n]$$

Matrix multiplications can be efficiently used during computation to obtain the overall transformation matrix.

Composite transformation will be illustrated through some solved problems at the end of this chapter.

3.3 GEOMETRIC TRANSFORMATIONS IN 3D

Geometric transformations like Rotation, Translation etc can be carried out in the 3D Cartesian space.

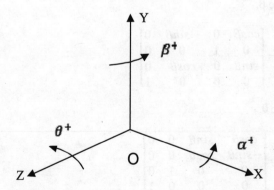

Figure 3.7 Right Handed 3D coordinate system

Figure 3.7 shows a right handed coordinate system, in which counter clockwise rotation about the coordinate axis is treated as positive.

Transformation matrices for various operations will be discussed.

3.3.1 Generalized Transformation Matrix

Using Homogeneous coordinate system, the generalized transformation equation is

P' = P [A]

[x' y' z' 1] = [x y z 1] [A]

Where **A** is the generalized affine transformation matrix

$$A = \begin{bmatrix} a11 & a12 & a13 & 0 \\ a21 & a22 & a23 & 0 \\ a31 & a32 & a33 & 0 \\ a41 & a42 & a43 & 1 \end{bmatrix} \qquad \textbf{3.8}$$

All the 3D transformations enumerated below will be the variants of [A].

Rotation

Figure 3.7 shows that a point can be rotated about the origin in 3D about X, Y or Z axis by angle α, β or θ respectively. As a result, three separate rotation matrices are required.

Rotation about X by angle α

$$R_x = \begin{bmatrix} 1 & 0 & 0 & 0 \\ 0 & \cos\alpha & \sin\alpha & 0 \\ 0 & -\sin\alpha & \cos\alpha & 0 \\ 0 & 0 & 0 & 1 \end{bmatrix} \qquad \textbf{3.9A}$$

Rotation about Y by angle β,

$$R_y = \begin{bmatrix} cos\beta & 0 & -sin\beta & 0 \\ 0 & 1 & 0 & 0 \\ sin\beta & 0 & cos\beta & 0 \\ 0 & 0 & 0 & 1 \end{bmatrix} \qquad \textbf{3.9B}$$

Rotation about Z by angle θ,

$$R_z = \begin{bmatrix} cos\theta & sin\theta & 0 & 0 \\ -sin\theta & cos\theta & 0 & 0 \\ 0 & 0 & 1 & 0 \\ 0 & 0 & 0 & 1 \end{bmatrix} \qquad \textbf{3.9C}$$

Note that the signs of terms are reversed in rotation matrix R_y as the rotation causes mapping from Z to X axis.

Scaling

The 3D scaling transformation matrix is given by

$$S = \begin{bmatrix} s_x & 0 & 0 & 0 \\ 0 & s_y & 0 & 0 \\ 0 & 0 & s_z & 0 \\ 0 & 0 & 0 & 1 \end{bmatrix}$$

3.10

where s_x, s_y, s_z are scaling factors in X, Y, Z direction about the origin.

Mirroring

Mirroring will cause image (reflection) of a point about X-Y, Y-Z, Z-X planes. For example to mirror about X-Y plane, the mirroring matrix will be

$$M = \begin{bmatrix} 1 & 0 & 0 & 0 \\ 0 & 1 & 0 & 0 \\ 0 & 0 & -1 & 0 \\ 0 & 0 & 0 & 1 \end{bmatrix}$$

3.11

Translation

To translate a point in 3D by [Δx, Δy, Δz] the transformation matrix will be

$$M = \begin{bmatrix} 1 & 0 & 0 & 0 \\ 0 & 1 & 0 & 0 \\ 0 & 0 & 1 & 0 \\ \Delta x & \Delta y & \Delta z & 1 \end{bmatrix}$$

3.12

3.3.2 Composite Transformations

It can be seen that all the geometric transformations discussed above (Equation 3.9-3.12) can be unified in the generalized transformation matrix **[A]** (Equation 3.8)

$$A = \left[\begin{array}{ccc:c} a11 & a12 & a13 & 0 \\ a21 & a22 & a23 & 0 \\ a31 & a32 & a33 & 0 \\ \hdashline a41 & a42 & a43 & 1 \end{array} \right]$$

The **[A]** matrix can be partitioned as above. It is seen that the upper 3x3 matrix (**a11-a33**) represents Rotation, Scaling and Mirroring transformations. The lower 1x3 matrix (**a41-a43**) represents Translation terms. The last column of **[A]** is always $[0\ 0\ 0\ 1]^T$ for affine transformations.

Similar to 2D transformation (section 3.2.2), composite transformation can be obtained in 3D space by applying series of transformations in succession

$$P' = P. [A_1] [A_2] [A_3] \dots \dots [A_n]$$

These will be illustrated through some examples

3.4 EXAMPLES

1. A unit square in X-Y plane with one vertex at origin is first rotated in counterclockwise direction by 30˚and then translated by a vector [5,5]. Compute the transformed position of the object. If the order of geometric transformations is reversed, will the positions of the transformed object vertices change? Justify your answer.

Solution:

The composite transformations equation will be,

$$P' = P[R][T]$$

$$[x'\ y'\ 1] = [x\ y\ 1] \begin{bmatrix} cos30° & sin30° & 0 \\ -sin30° & cos30° & 0 \\ 0 & 0 & 1 \end{bmatrix} \begin{bmatrix} 1 & 0 & 0 \\ 0 & 1 & 0 \\ 5 & 5 & 1 \end{bmatrix}$$

Multiplying out the transformation equations will be

$$X' = x\ cos30˚- y\ sin30˚+ 5$$
$$Y' = x\ sin30˚+ y\ cos30˚+ 5$$

Using these equations, the vertices of the object $P_1[0, 0]$, $P_2[1, 0]$, $P_3[1, 1]$, $P_4[0, 1]$ can be transformed. The transformed vertices of the object will be $P_1'[5,5]$, $P_2'[5.866,5.5]$, $P_3'[5.366,6.366]$, $P_4'[4.5,5.866]$.

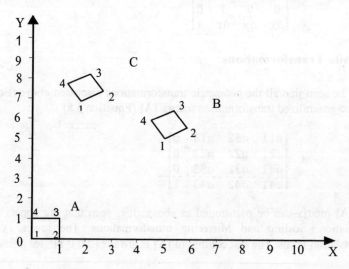

Figure 3.8 Composite Transformation

In Figure 3.8, A, B show the original and the transformed objects

If the order of transformation is reversed, the transformation equation is

$$P' = P [T] [R]$$

Proceeding as earlier, the new transferred vertices will be

$P_1'[1.830, 6.830]$, $P_2'[2.696, 7.330]$, $P_3'[2.196, 8.196]$, $P_4'[1.33, 7.696]$

In Figure 3.8, C shows the object transformed under this condition which is different than the object B.

This is because the matrix multiplication is not commutative.

2. Compute the image of the point **P[6,7]** about a line passing through the points $P_1[1,1]$ and P_2 **[10,8].** Verify your answer.

Solution:

Figure 3.9 shows the line P_1P_2 and the point **P.**

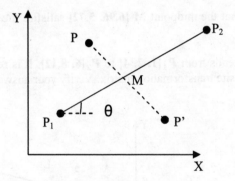

Figure 3.9 Mirroring about a Line

Since the image is to be obtained about the line P_1, P_2 (Figure 3.9), series of transformations need to be done to map to the standard conditions discussed earlier.

- Translate P_1 to origin
- Rotate clockwise by angle θ
- Mirror about X-axis
- Rotate counterclockwise by angle θ
- Translate P_1 back to its original position

The composite transformation equation is

$$P' = P\ [T_1]\ [R_1]\ [M]\ [R_2]\ [T_2]$$

Where
$$T_1 = \begin{bmatrix} 1 & 0 & 0 \\ 0 & 1 & 0 \\ -2 & -2 & 1 \end{bmatrix},\ R_1 = \begin{bmatrix} 0.8 & -0.6 & 0 \\ 0.6 & 0.8 & 0 \\ 0 & 0 & 1 \end{bmatrix},\ M = \begin{bmatrix} 1 & 0 & 0 \\ 0 & -1 & 0 \\ 0 & 0 & 1 \end{bmatrix}$$

$$T_2 = \begin{bmatrix} 1 & 0 & 0 \\ 0 & 1 & 0 \\ 2 & 2 & 1 \end{bmatrix},\ R_2 = \begin{bmatrix} 0.8 & -0.6 & 0 \\ 0.6 & 0.8 & 0 \\ 0 & 0 & 1 \end{bmatrix}$$

Solving the above equation, the transformation equations are

$$X' = 0.28x + 0.96y - 0.48$$
$$Y' = 0.96x - 0.28y + 0.64$$

Using these equations, the image of **P [6, 7]** is **P' [7.92, 4.44]**. If P' is the image of P, line P_1P_2 will bisect line **PP'**. Midpoint of line PP' is **M [6.96, 5.72]**.

From Figure 3.9, the equation of line $P_1 P_2$ is

y - 0.75x - 0.5 = 0

It can be seen that the midpoint **M [6.96, 5.72]** satisfies the above equation. Hence **P'** is the image of **P**.

3. A ray in 3D space extends from $P_1[1, 2, 4]$ to $P_2[6, 8, 12]$. It is required to align this ray with the +Z-axis. Derive the requisite transformation matrix. Verify your answer.

Solution:

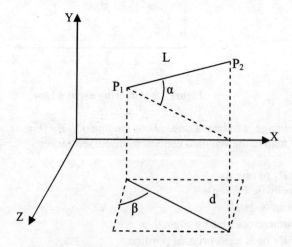

Figure 3.10 Rotation in 3D

Figure 3.10 show the ray $P_1 P_2$ in the 3D space

To align the ray along + Z-axis the following transformations will be required, in order

- Translate P_1 to origin
- Rotate clockwise about **Y** by angle **β**
- Rotate counterclockwise about **X** by angle **α**

The composite transformation matrix will be

$$P' = P [T_1] [R_y] [R_x]$$ **3.13**

To compute angles α and β.

The direction ratios (DRs) of the ray $P_1 P_2$ are **[5, 6, 8]** (Figure 3.10)
Length of Line L $=\sqrt{(5^2 + 6^2 + 8^2)} = 11.18$
Length of the projection of line on X-Z plane (d) $=\sqrt{5^2 + 8^2}$
$$= 9.434$$

Using L and d (Figure 3.10)

Cos β= 0.848, sin β= -0.530

Cos α=0.844, sin α= 0.537

Substituting in Equation 3.11, the final transformation equations are

X'= 0.848x – 0.53z +1.272
Y'=-0.2846x + 0.844y -0.4554z+0.4181
Z'= 0.4473x + 0.5370y + 0.7157z – 4.3842

The above equations can be verified by applying to P_1 and P_2. The transformed points are P'_1[0, 0, 0, 1] and P'_2 [0, 0, 11.18, 1].

It shows that the transformed ray $P'_1 P'_2$ lies along + Z-axis

3.5 REVIEW QUESTIONS

1. What are Euclidean geometric transformations? How do they differ from the non Euclidean ones? Give examples.

2. Explain in brief, why Homogeneous Coordinate System is preferred for carrying geometric transformations in Computer Graphics?

3. In a Computer aided drafting package, a Triangle and a Square have been created. The triangle has vertices **A [0,0]**, **B [5.0]**, **C [0,5]** and the square has vertices P_1 **[5,5]**, P_2 **[8,5]**, P_3 **[8,8]** and P_4 **[5,8]**. Derive requisite transformation matrix to position vertex P_1 to coincide with **B** and edge P_1P_4 to coincide with **BC**.

4. During Shear transformation, the vertices of a square object P_1 [0,0], P_2 [1.0], P_3 [1,1], P_4 [0,1] get transformed to P'_1 [0,0], P'_2 [2,1], P'_3 [3,3], P'_4 [1,2]. Computer the shear transformation matrix.

5. A tetrahedron with vertices P_1 [0,0,0], P_2 [1,0,0], P_3 [0,1,0] and P_4 [0,0,1] is first rotated about **Z** axis by 20 degrees in clockwise direction, followed by rotation about **X** axis by 10 degrees in counterclockwise direction. The object is translated by a vector **[8,8,8]**. Compute the vertices of the object after applying all these transformations.

Chapter 4

Projection Transformations

Projection is a transformation operation in which the geometry of a 3 D object is mapped into 2 D coordinate space for representation on paper or the Computer monitor. It is an important operation in Computer graphics, geometric modeling and computer gaming. Various types of transformations exist to suit the end applications of the projected image.

This chapter will present in details, the mathematical basis of the projection transformations with suitable examples.

4.1 FROM OBJECT TO IMAGE

Similar to the geometric transformations, topology of the object does not change under the projection transformation. The fundamental geometric entity undergoing transformation is the vertex of the object represented as a position vector **P[x, y, z]** in $\mathbf{R}^3$. Projection is, in essence, mapping from 3D object space to 2D image space.

The process of projection can be conceptually conceived as taking the snapshot of an object using a camera. In computer graphics, a virtual 3D camera is used to accomplish the task of projection. Figure 4.1 shows the right handed world coordinate system (WCS) $\mathbf{X_W}$-$\mathbf{Y_W}$-$\mathbf{Z_W}$ in which the object is modeled. User defines the virtual 3D camera at the chosen point in WCS termed as View Reference Point **(VRP) P_0 [x_0 y_0 z_0]**. A view plane **(VP)** designated by $\mathbf{X_v}$ –$\mathbf{Y_v}$– $\mathbf{Z_v}$ is set up in WCS to *look* at the object in a desired orientation. Projector rays emanating from the vertices of the object intersect the view plane to create projection of the object.

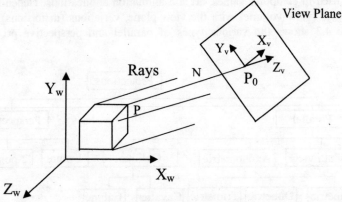

Figure 4.1 Setting up 3D Virtual Camera.

4.1.1 Types of Projections

Virtual 3D camera enables the user to obtain different projections of the object depending upon the end application of the projected image. For a chosen camera position and orientation **(VRP and VP)**, the type of projection is governed by shape of view volume decided by the rays.

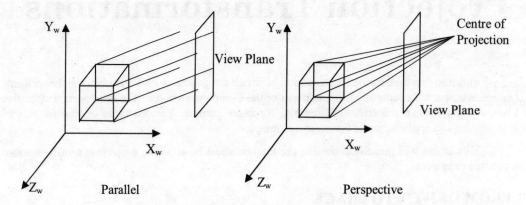

Figure 4.2 Parallel and Perspective Projections

Figure 4.2 shows the view volumes for Parallel and Perspective projections. In parallel projection, the rays emanating from the vertices are parallel to each other (converge at infinity) while in perspective projections, they converge at a finite point called as Center of Projection **(COP)**. The view volumes constituted by the rays are thus, a parallelepiped for parallel projection and a Truncated Pyramid for perspective projection.

Parallel projections are most widely used by Engineers for computer aided drafting and modeling as they do not introduce any distortions in dimensions, angles and can thus, be used for measurement. They however, lack realism in the image. Perspective projections on the other hand, provide realism in image but introduce length and angle distortions in the image. Perspective projection is widely used in computer games, art and animation applications. Depending on the angle of incidence of the rays (view volume) with the view plane, variations (distortions) in the projected image occur. Figure 4.3 shows the various types of parallel and perspective projections used in practice.

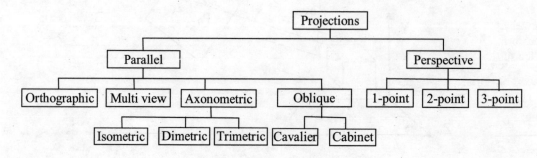

Figure 4.3 Types of Projections

Mathematical basis for obtaining various types of projections will be presented in the sections to follow.

4.2 MATHEMATICS OF PROJECTION

4.2.1 Orienting 3 D Camera

To obtain projection of the object on the view plane, the 3D Virtual camera needs to be properly positioned and oriented with reference to the object (world) coordinate System (WCS).

Figure 4.1 shows the WCS (X_w, Y_w, Z_w) and the object to be projected A view Coordinate system (X_v, Y_v, Z_v) is attached to the camera. Steps to locate and orient the 3D camera in WCS are enumerated below.

- Define View Reference Point (VRP) $P_0(x_0, y_0, z_0)$ for locating the camera
- Choose a convenient point P in the world to *look* at. It could be a point on the object which is the focus of view for the camera.
- Define view direction PP_0 and obtain normal vector $N = \dfrac{PP_0}{|PP_0|}$
- Define the View Coordinate system (X_v, Y_v, Z_v) on the view plane such that Z_v is along N. Axes X_v, Y_v lie in the View plane which is orthogonal to **N**.
- Define an up vector (**UV**) to set orientation of the camera (X_v, Y_v) Any convenient vector can be used as up vector UV e.g. Y_w .
- Compute X_v, Y_v unit vectors passing through P_0 and lying in the viewing plane
- The camera is now fully defined in terms of $P_0,$ X_v, Y_v, Z_v.

It is simple and efficient to compute the projection of the object on X_w -Y_w plane. As a result, without loss generality, the View Plane along with the associated world can be transformed to align it with X_w -Y_w plane.

The camera is transformed along with the objects to the origin of WCS using series of geometric transformations (Rotation, Translation) so that P_0 lies at the origin and $[X_v, Y_v, Z_v]$ aligns with $[X_w, Y_w, Z_w]$. This can be done using geometric transformation discussed in Chapter 3.

4.2.2 Projection Vector and Plane Equations

In subsequent dissuasions, without loss of generality, it will be considered that the camera view system $[X_v, Y_v, Z_v]$ is positioned and aligned with WCS $[X_w, Y_w, Z_w]$ and the object(s) are transformed accordingly. As a result, the subscript w will be dropped from WCS $[X_w, Y_w, Z_w]$ notation in the subsequent discussions.

To obtain the projection of a vertex (point) $P_1[x_1, y_1, z_1]$, one needs to compute the intersection of a ray (projector) passing through P_1 with the projection plane X-Y (z = 0). Figure 4.4 shows the point P_1, the projected point P_p, the projector ray and X-Y plane.

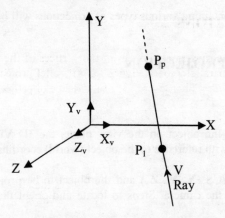

Figure 4.4 Projection of a Point on Plane

Let the projector ray be defined as a vector **V** (V_x, V_y, V_z). The vector valued parametric equation of the projector (ray) can be written as

$$P(u) = P_1 + uV \qquad \qquad 4.1$$

where u is a scalar. By assigning various values to u, different points on the projector (ray) can be computed. The projector ray (Equation. 4.1) will intersect with the X-Y plane (z=0) for u= u* where

$$u^* = - Z_1 / V_z \qquad \qquad 4.2$$

Substituting Equation 4.2 in Equation 4.1, the projected coordinates are

$$X_p = X(u^*) = X_1 - (V_x / V_z)Z_1$$
$$Y_p = Y(u^*) = Y_1 - (V_y / V_z)Z_1$$
$$Z_p = 0 \qquad \qquad 4.3$$

Generalized projection equation can be written as

$$P_p = P_1 \,[PM]$$
$$[x_p \; y_p \; z_p \; 1] = [x_1 \; y_1 \; z_1 \; 1] \; [PM] \qquad \qquad 4.4$$

where **P$_p$** denotes projected point. The projection transformation matrix **[PM]** is given by

$$[PM] = \begin{bmatrix} 1 & 0 & 0 & 0 \\ 0 & 1 & 0 & 0 \\ -\dfrac{v_x}{v_z} & \dfrac{-v_y}{v_z} & 0 & 0 \\ 0 & 0 & 0 & 1 \end{bmatrix}$$

Equation 4.4 is the basic projection equation applicable to both Parallel and Perspective projections. Specific cases shown in Figure 4.3 are discussed one by one.

4.3 PARALLEL PROJECTIONS

In parallel projection, the rays emanating from the vertices of the object are parallel to each other and the view volume is a parallelopiped (Figure 4.2). Parallel projection can be classified into **Orthographic or Oblique** depending upon the angle of incidence of projector ray with the projection plane.

4.3.1 Orthographic Projections

In orthographic projections, the view volume (rays) intersects the projection plane orthogonally. The projected image has true lengths and angles as the object and can thus, be used for measurement. Orthographic projections to get Front, Top, Side views of objects are standardized in Engineering Drawing Practice and are widely used by Engineers worldwide

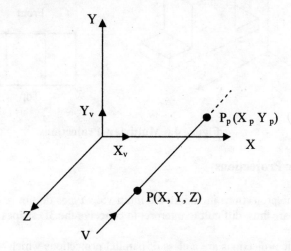

Figure 4.5 Orthographic Projection of a Point

Figure 4.5 shows the orthographic projection of a point **P[x, y, z]** on X-Y plane. The view vector is **V [0, 0, 1]**. The projected point **P_p** is

$$P_p = P \, [PM]$$
$$[x_p \; y_p \; z_p \; 1] = [x \; y \; z \; 1] \, [PM]$$

where the projection matrix **[PM]** is given by

$$[PM] = \begin{bmatrix} 1 & 0 & 0 & 0 \\ 0 & 1 & 0 & 0 \\ 0 & 0 & 0 & 0 \\ 0 & 0 & 0 & 1 \end{bmatrix} \qquad \textbf{4.5}$$

Equation 4.5 shows that for orthographic projection, the Z coordinates of the object vertices are simply dropped

4.3.2 Multi View Projections

Orthographic projections give *true* dimensions of the objects but do not give any idea about its 3D shape. As a result, Multi view projections are used in the Engineering drawing practice.

Multi view projections are in essence, the orthographic projections of the object on the X-Y, Y-Z, Z-X planes. These are termed as Front, Side and Top views. Figure 4.6 shows an object and its multi views. These can be computed by locating and orienting the 3D camera in world (WCS) and computing the projection matrix accordingly.

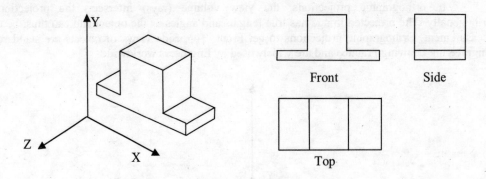

Figure 4.6 Multiview Projections

4.3.3 Axonometric Projections

Orthographic projections including the multi view types do not really show multiple faces of the object in 3D and are thus, difficult to interpret to conceive the 3D shapes

Axonometric projections are a class of parallel projections which show at least 3 faces of the object in the projected view. The 3D camera is oriented properly in world. The projector rays from the object always intersect the view plane orthogonally. The projected edges in the image are foreshortened depending upon the inclination of the view plane (view vector N or V) with the X-Y-Z axes. Three cases arise as under,

Isometric Projection – View vector equally inclined to X-Y-Z axes.
Dimetric Projection -   View vector equally inclined to two axes (generally X-Z).
Trimetric Projection -   View vector makes unequal angles with X, Y, Z axes.

Figure 4.7 A, B show the Isometric and Dimetric projections of a unit cube. Isometric projections are most widely used in computer games and animation to get 3D view of scenes/world. Dimetric projection is particularly used in engineering practice to represent objects to bring in measurement as well as 3D shape view. Trimetric is the least restrictive projection and is thus, the general case of axonometric projections. However due to the difficulties in visualization and lack of standardization of proportions, it is not preferred compared to the Isometric and Dimetric types of projections.

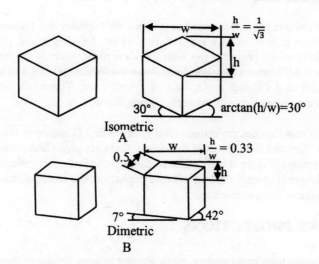

Figure 4.7 Isometric and Dimetric Projections

To get proper projection matrices, user can choose the View vector appropriately depending upon the intended need of projections viz. the isometric, dimetric or trimetric projection.

4.3.4 Oblique Projections

Oblique projections are the most general class of parallel projections. The projected rays (parallelopiped) from the object intersect the view plane in an inclined, non-orthogonal manner. No restrictions exist on the orientation of the view plane normal (N) and the projected ray direction (V).

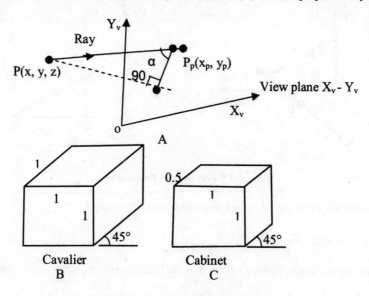

Figure 4.8 Oblique Projections

Two popular oblique projections used in practice are Cavalier and Cabinet projections. Figure 4.8 A shows the oblique projection of a point **P [x y z]** on the view plane **[Xv Yv]**. The angle α indicating the inclination of the projector ray with the view plane governs the type of projection. For Cavalier projection, $\alpha = 45°$ (tan-1 (1)) while for Cabinet projection, $\alpha = 63.4°$ (tan-1 (2)). Figure. 4.8 B C show the Cavalier and Cabinet projections of a unit cube. These can be produced by using projection matrices discussed earlier with properly chosen N and V.

Both Cavalier and Cabinet projections tend to bring in 3 D nature in the image by displaying combination of Front, Top and Side views of the object. Cavalier projection preserves lengths of lines perpendicular to view plane (Figure 4.8 B) while in cabinet one lines perpendicular to the view plane project at ½ of their length (Figure 4.8 C). In cavalier projection, the shape is distorted. Comparatively cabinet projection produces more realistic view.

4.4 PERSPECTIVE PROJECTIONS

Perspective projections bring realism by producing images similar to those seen by the human eye. Objects farther from the eye look smaller and nearer ones look bigger. The projector rays from the object converge to point termed as Center of Projection (COP) which is similar to human eye. The view volume constituted by the projectors is thus, a cone/pyramid (Figure 4.2)

4.4.1 Theory of Perspective Projections

Figure 4.9 A shows the perspective projection of a point $P_1[x_1, y_1, z_1]$ on the X-Y plane. The projector ray passes through the centre of projection $P_c [x_c, y_c, z_c]$, the point P_1 and intersects the X-Y plane to give projection $P_p [x_p, y_p, z_p]$.

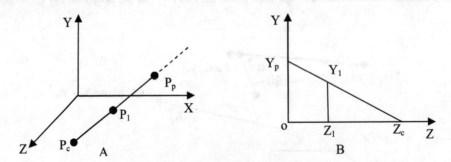

Figure 4.9 Perspective Projection

Vector valued parametric equation of the projector can be written as

$$P(u) = P_c + u(P_1 - P_c) \qquad\qquad 4.6$$

where u is a scalar. By assigning various values to u, different points on the projector can be computed.

The projector intersects the X-Y plane at a specific value of $u = u^*$ given by

$$u^* = - z_c / (z_1 - z_c)$$

Substituting this value of u* into Equation 4.6, the projected point is obtained as

$$X_p = X_c + [(X_1 - X_c) / (1 - (z_1 / z_c))]$$
$$Y_p = Y_c + [(Y_1 - Y_c) / (1 - (z_1 / z_c))]$$
$$Z_p = 0 \qquad\qquad\qquad\qquad\qquad\qquad 4.7$$

To check if Equation 4.7 provides the perspective projection of point P_1 as P_p. Without loss of generality, the centre of projection is chosen as P_c [0, 0, z_c] on the Z axis. Figure 4.9 B shows the projection of P_1 in the Y-Z plane.

Considering similar triangles,

$$Y_p / Y_1 = z_c / (z_c - z_1)$$
$$Y_p = Y_1 / (1 - (z_1 / z_c))$$

Similarly $X_p = X_1 / (1 - (z_1 / z_c))$ $\qquad\qquad\qquad\qquad$ **4.8**

Equation 4.7 gives the same equations for perspective projections for P_c [0, 0, z_c].

4.4.2 Single and Multi Point Projections

Projection transformations bring realism by distorting the image. Parallel lines on the object do not remain parallel in the image.

Depending upon the inclination of the view plane with the object, three types of perspective projections occur viz 1- Point, 2-Point and the 3-Point projections.

Single Point Perspective Projections

In Single point perspective projections, the view plane is parallel to the X-Y plane and intersects the Z axis. Using suitable geometric transformations, without loss of generality it is considered that the view plane (projection plane) is aligned with the X-Y plane and the centre of projection is P_c [0, 0, z_c].

Figure 4.10 A, B shows a cube of size 2x2x2 and its Single point perspective projection with respect to P_c [0, 0, 10]. The projection equations in Equation 4.7/4.8 can be represented in Homogeneous coordinate system as under

$$P_p = P [PM]$$
$$[x_p\ y_p\ z_p\ 1] = [x\ y\ z\ 1]\ [PM]$$

The projection matrix [PM] is given by

$$[PM] = \begin{bmatrix} 1 & 0 & 0 & 0 \\ 0 & 1 & 0 & 0 \\ 0 & 0 & 0 & -1/zc \\ 0 & 0 & 0 & 1 \end{bmatrix} \qquad\qquad 4.9$$

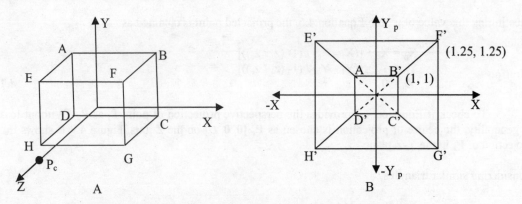

Figure 4.10 Single Point Perspective Projection of a cube

Important observations from Figure 4.10 are as under

- Face E'F'G'H' (Figure 4.10 B) looks larger than face A'B'C'D', though they both are of the same size. Face E'F'G'H' is closer to P_c than face A'B'C'D'

- Parallel line (edges) such as AE, BF etc on the object do not remain parallel in image (A'E', B'F')

- These edges when extended appear to converge at origin (Figure 4.10 B) or at a point on the Z axis. This is termed as the *Vanishing Point* (VP). The projection is termed as single point as there is a Single Vanishing Point. In the present case; $VP_z = -zc$.

Single point perspective projections are commonly observed by human eye. Parallel railway tracks or long highways appear to be converging. The walls of tunnel appear trapezoidal and converging. Single point perspective projection is widely used in the works of art.

Figure 4.11 Painting using single point perspective projection
{Ref:http://en.wikipedia.org/wiki/File:Piazza_San_Marco_with_the_Basilica,_by_Canaletto,_1730._Fogg_A rt_Museum,_Cambridge.jpg}

Figure 4.11 shows the famous painting Piazza San Marco with the Basilica by Canaletto, 1730 which uses single point perspective projection.

Multipoint Perspective Projection

The 2-point and 3-point perspective projections basically indicate that there are 2 or 3 vanishing points. Figure 4.12 shows the 2-point and 3-point perspective projection of a cube. 2-point projection occurs when view plane is parallel to Y-axis and cuts X and Z axes. In 3-point projections, the view plane cuts all the three axes. These are more widely used in the architectural drawings for buildings.

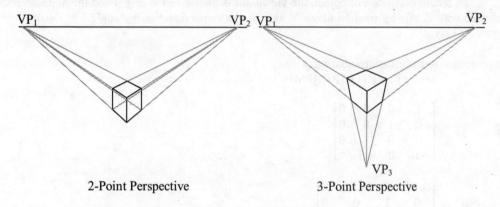

2-Point Perspective   3-Point Perspective

Figure 4.12 Multipoint Perspective Projections

4.4.3 Generalized Projection Transformation Matrix

Generalized perspective projection transformation matrix can be obtained by carrying out geometric transformations to orient 3D camera, object etc and coincide view plane with X-Y.

$$P_p = P \, [G] \, [PM]$$

where **[G]** is the overall geometric transformation matrix. Generalized perspective projection transformation matrix **[PM]** is given by

$$[PM] = \begin{bmatrix} 1 & 0 & 0 & p \\ 0 & 1 & 0 & q \\ 0 & 0 & 0 & r \\ 0 & 0 & 0 & 1 \end{bmatrix}$$ **4.10**

$p = 1/vp_x$, $q = 1/vp_y$, $r = 1/vp_z$

Where vp_x, vp_y, vp_z are the coordinates of vanishing points on the X,Y,Z axes.

It can be seen that the projection matrix **[PM]** in Equation 4.10 is not Affine ($p, q, r \neq 0$) which introduces distortions (foreshortening) in the image.

4.5 EXAMPLES

1. Compute parallel projection of an object shown in Figure. 4.13 on a view plane passing through the point P_0 [10,0,5] and having the view vector in the direction O-P_0.

Solution:

View plane normal vector $N = \frac{OP_0}{|OP_0|} = [\,0.89944,\ 0,\ 0.4472\,]$

To obtain projection of object, the viewpoint is transferred to origin and the normal vector N is aligned with Z axis by rotation about Y axis in clockwise direction by angle β = -70.48 degrees. (Figure 4.13 A)

The projection transformation equation is
$$P_p = P\,[T]\,[R_y]\,[PM]$$

where $\quad T = \begin{bmatrix} 1 & 0 & 0 & 0 \\ 0 & 1 & 0 & 0 \\ 0 & 0 & 1 & 0 \\ -10 & 0 & -5 & 1 \end{bmatrix}$

$$R_y = \begin{bmatrix} \cos\beta & 0 & -\sin\beta & 0 \\ 0 & 1 & 0 & 0 \\ \sin\beta & 0 & \cos\beta & 0 \\ 0 & 0 & 0 & 1 \end{bmatrix}$$

$$PM = \begin{bmatrix} 1 & 0 & 0 & 0 \\ 0 & 1 & 0 & 0 \\ 0 & 0 & 0 & 0 \\ 0 & 0 & 0 & 1 \end{bmatrix}$$

Using the above equation, parallel projection of the object is computed. Table below shows the original and projected coordinates of the vertices of the object.

No	Object Coordinates			Projected Image	
	X	Y	Z	X_p	Y_p
1	0	0	0	0	0
2	1	0	0	0.4772	0
3	1	1	0	0.4772	1.0
4	0	1	0	0	1.0
5	1	1	0.5	0	1.0
6	0	0	1	-0.8994	0
7	1	0	1	-0.4772	0
8	1	0.5	1	-0.4772	0.5
9	0.5	1	1	-0.6708	1.0
10	0	1	1	-0.8944	1.0

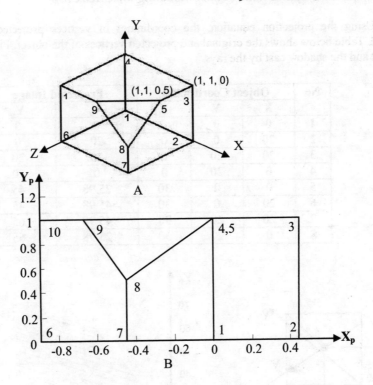

Figure 4.13 Object and its Projection

The projected image (Figure 4.13 B) shows that the edges of the object are foreshortened indicating that it is an axonometric projection.

2. Ray of the sun fall on a building having a base of 20 x 20 meters and height of 30 meters. The rays make angles of 64.34, 41.4 and 120 degrees with X, Y and Z axes respectively. Compute the shadow of the building on the ground (X-Y).

Solution:

Rays of the sun are parallel and fall at an angle on the ground. This is a case of oblique parallel projection.

The ray vector in terms of the direction cosines is **V [0.4330, 0.75, -0.5].**The projection equation becomes

$$P_p = P \; [PM]$$

where the projection matrix [PM

$$PM = \begin{bmatrix} 1 & 0 & 0 & 0 \\ 0 & 1 & 0 & 0 \\ -\dfrac{v_x}{v_z} & -\dfrac{v_y}{v_z} & 0 & 0 \\ 0 & 0 & 0 & 1 \end{bmatrix}$$

Using the projection equation, the coordinates of vertices projected on X-Y plane are computed. Table below shows the original and projected vertices of the object. Figure 4.14 A, B shows the object and the shadow cast by the rays.

No	Object Coordinates			Projected Image	
	X	Y	Z	X_p	Y_p
1	0	0	0	0	0
2	20	0	0	20	0
3	20	20	0	20	20
4	0	20	0	0	20
5	0	0	30	25.98	45
6	20	0	30	45.98	45
7	20	20	30	45.98	65
8	0	20	30	25.98	65

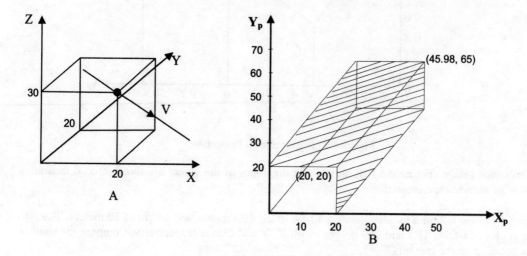

Figure 4.14 The Building and its Shadow

3. A unit cube is first rotated about Y axis in a clockwise direction by 60 degrees followed by rotation about X axis in the counter clockwise direction by 30 degrees. Compute single point perspective projection of the rotated cube on X-Y plane with center of projection P_c [0,0,4].

Solution:

The projection equation will be as under

$$P_p = P [R_y] [R_x] [PM]$$

where

$$R_y = \begin{bmatrix} \cos\beta & 0 & -\sin\beta & 0 \\ 0 & 1 & 0 & 0 \\ \sin\beta & 0 & \cos\beta & 0 \\ 0 & 0 & 0 & 1 \end{bmatrix} \qquad \beta = -60°$$

$$R_x = \begin{bmatrix} 1 & 0 & 0 & 0 \\ 0 & \cos\alpha & \sin\alpha & 0 \\ 0 & -\sin\alpha & \cos\alpha & 0 \\ 0 & 0 & 0 & 1 \end{bmatrix} \qquad \alpha = 30°$$

$$PM = \begin{bmatrix} 1 & 0 & 0 & 0 \\ 0 & 1 & 0 & 0 \\ 0 & 0 & 0 & \dfrac{-1}{z_c} \\ 0 & 0 & 0 & 1 \end{bmatrix} \qquad Z_c = 4$$

Using the above equation, vertices of the object are transformed. Table below shows the original and the projected coordinates.

No	Object Coordinates			Projected Image	
	X	Y	Z	X_p	Y_p
1	0	0	0	0	0
2	1	0	0	0.6156	-0.5335
3	1	1	0	0.7276	0.6306
4	0	1	0	0	0.9905
5	0	0	1	-0.9720	-0.2804
6	1	0	1	-0.5210	-0.9708
7	1	1	1	-0.6335	0.3167
8	0	1	1	-1.1305	0.8044

Figure 4.15 shows the projected image of the object. It can be seen the image is a 3– point perspective projection as it has three vanishing points

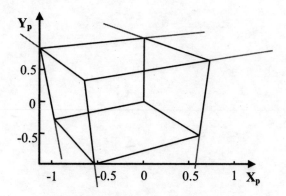

Figure 4.15 Projection of the Solid

4.6 REVIEW QUESTIONS

1. Bring out the difference between an axonometric projection and a perspective projection.

2. State *True or False* with reasons.
   a.   A perspective projection is an irreversible transformation.
   b.   Trimetric projection is a form of 3 point perspective projection.

3. Derive projection transformation matrices to obtain Front, Top and Side views of an object for creating an engineering drawing.

4. Compute the isometric projection of an object shown in Figure. 4.13 on a view plane passing through view reference point P_0 [10,10,10].

5. A unit cube is first rotated about Y axis in a clockwise direction by 30 degrees followed by translation by vector [0,-5,0] Compute single point perspective projection of this transformed cube from the centre of projection P_c [0,0,10]. Does the transformed image have multiple vanishing points? Justify your answer.

Chapter **5**

Geometric Design of Planar and Space Curves

Curves have fascinated mankind since ages. This is manifested in the works of art, sculpture, jewellary, textile designs, domestic and industrial products. Curves form an important geometric element in the design of products. In computer aided geometric design (CAGD) applications, designers need interactive tools for the synthesis of complex curves and surfaces to create aesthetically smooth products.

This chapter presents in details, the mathematical techniques for the design- synthesis of various planar and space curves used in CAGD. Suitable examples have also been included.

5.1 TYPES OF CURVES

Curves can broadly be classified into planar and space curves. Planar curves lie completely in a plane and are essentially 2D in nature. Figure 5.1 shows standard analytical planar curves like Circle, Ellipse, Hyperbola and Parabola which are obtained by the intersection of a plane with the cone. Space curves, on the other hand, do not lie in a plane and thus, have the ability to twist in space.

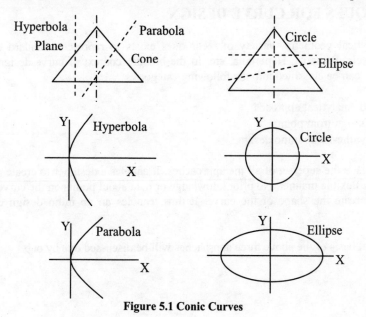

Figure 5.1 Conic Curves

Curves can further be classified as Open or Closed curves based on the nature of the curve. For example a trigonometric SIN or COSINE function curve is Open while a Circle /Ellipse is a closed curve. Some curves loop around and intersect with themselves. In computer aided geometric design, such self intersecting curves are not used in product design. They will be out of scope of this chapter.

5.2 BASIS FOR CURVE REPRESENTATION

Curves have been traditionally drawn on paper or represented in terms of points on the curve. In essence, a planar or space curve can be conceived as the locus of a point which moves in $\mathbf{R^2}$ $(\mathbf{R^3})$ space according to some predefined law or equation. The extent of movement of the point determines the bounds of the resulting curve.

Though the representation of curve in terms of points on the curve is simple and straight forward, it is beset with several limitations. It is difficult to compute intermediate points on the curve and one may need to use some numerical interpolation technique (Newton, Lagrange).Further the properties of the curve such as length, Tangent, Normal, curvature cannot be easily computed and may again need some numerical procedures. These properties are particularly important in CAGD to manipulate the shape of the curve and blend it to another curve segment to achieve continuity. From memory consideration, curve representation in the terms of points will be quite verbose and inefficient

As a result, planar and space curves are represented in terms of curve equations. The functions which represent the curve equations could be polynomial (Linear, Quadratic, and Cubic), Trigonometric, Power, Exponential or some complex basis functions depending upon the complexity of the curve being represented.

Techniques for curve design and the associated forms of mathematical functions will be discussed in the sections to follow.

5.3 TECHNIQUES FOR CURVE DESIGN

In analytical geometry, variety of techniques exists to represent standard curves such as Straight line, Circle, Parabola, Hyperbola, etc. In the broader context of curve design/ synthesis for CAD/CAM, they can be classified into the following categories

- Standard Analytical approach
- Curve Design from points
- Curve synthesis- ab initio design

Synthesis is the superset of all the approaches. It enables a designer to create and manipulate curve shapes in a flexible manner. No prior knowledge or data about points on the curve is required by the designer to create the shape of the curve. It thus, enables an ab initio design environment in CAD/CAM.

Mathematical basis of the above three approaches will be discussed one by one.

5.4 MATHEMATICAL BASIS FOR CURVE REPRESENTATION

5.4.1 Implicit and Explicit curve equations

In analytical geometry, a curve is represented by an equation specifying the relation between the coordinates X, Y, Z which determine the shape of the curve. Two forms of equations exist viz the *Explicit* and the *Implicit*. For example, a 2D planar curve is represented in the explicit equation form as

$$y = f(x)$$

The mapping function **f ()** could be polynomial, exponential, power, trigonometric etc based on the shape of the curve desired. Explicit curve equations are used to represent Open, Single valued functions. For example, a straight line is represented by a linear explicit equation of the form

$$y = a_0 + a_1x \qquad\qquad 5.1$$

Implicit equations are used to represent Closed, multi valued functions which represent conic sections like circle, ellipse, parabola, and hyperbola. The implicit curve equation is of the form

$$f(x, y) = 0$$

A general implicit equation to represent quadratic (degree two) conic curves is as under

$$Ax^2 + 2Bxy + Cy^2 + 2Dx + 2Ey + F = 0 \qquad\qquad 5.2$$

In matrix notation, the equation can be written as

$$XQX^T = 0, \text{ where } X = [x\ y\ 1], \quad Q = \begin{bmatrix} A & B & D \\ B & C & E \\ D & E & F \end{bmatrix}$$

In equation 5.2, the coefficient of **xy, x** and **y** are **2B, 2D** and **2E** respectively. The polynomial has six coefficients which could be reduced to five by dividing by a non zero one. This implies that five conditions can uniquely determine the types of the conic curve. In particular,
- if $B^2 - AC < 0$, the curve is an *Ellipse*
- if $B^2 - AC = 0$, the curve is a *Parabola*
- if $B^2 - AC > 0$, the curve is a *Hyperbola*

Using the general equation, the equation of a circle with centre **(a, b)** and radius **r** can be written as
$$(x - a)^2 + (y - b)^2 = r^2$$
$$\text{or}$$
$$x^2 + y^2 - 2ax - 2by + a^2 + b^2 - r^2 = 0 \qquad\qquad 5.3$$

In a similar way, the equation for ellipse with center at origin and semi major and semi minor axes as **a** and **b** will be

$$\frac{x^2}{a^2} + \frac{y^2}{b^2} = 1.0 \qquad\qquad 5.4$$

Implicit equation for the normal form of a hyperbola is

$$\frac{x^2}{a^2} - \frac{y^2}{b^2} = 1.0$$
5.5

Figure 5.1 shows the typical conic sections referred in Equation 5.3-5.5.

5.4.2 Visual representation of curves

In computer graphics, curves are represented in a piecewise linear fashion. Points on the curve are computed and joined by short segments of straight lines. Shape (quality) of the curve is, thus, dictated by the density and distribution of points on the curve. This can be illustrated by taking an example. It is desired to represent the arc of a circle with centre at origin and having unit radius in the first quadrant. The Cartesian equation of the circle is

$$x^2 + y^2 = 1$$
5.6

Points on the arc of circle are computed by taking points along X axis in equal intervals viz X= [0, 0.2, 0.4, 0.6, 0.8, 1.0]. Using Equation 5.6, corresponding y values are computed. Figure 5.2 shows the arc of the circle and its piecewise approximation obtained from the computed points.

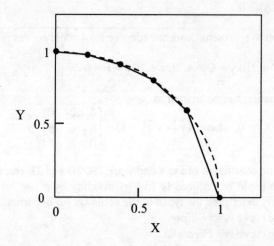

Figure 5.2 Arc of circle and piecewise approximation

It can be seen that the computed points provide a poor visual representation of the circle. Following observations can be made from Figure 5.2.

- Equal increments along X axis have not led to equal increments along Y.
- More number of points have got bunched on the top part of curve (around Y=1.0) compared to the regions of large slope (vertical part). The slope of the curves is continuously varying from very low to very high values.
- Length of arc along the curve between successive points is varying.

The Cartesian implicit form of equation with the above computing scheme is unable to provide a good distribution of points, particularly in the zones of high slopes. The distribution of points along the curve is not uniform. A similar problem would arise for curve shapes with varying degrees of curvature. Regions of smaller curvatures would need larger point density to improve the quality of representation (Figure 5.3)

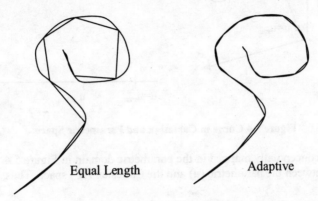

Equal Length   Adaptive

Figure 5.3 Point Density and Arc Length

This difficulty is offset to a large extent by using Parametric from of curve representation instead of the Cartesian implicit/explicit form.

5.4.3 Vector Valued Parametric Equations

In the Cartesian implicit/explicit curve equations, the Cartesian coordinate of points on the curve are expressed as equations of the form $y = f(x)$ or $g(x, y) = 0$. These forms lead to problems discussed earlier. A major difficulty is the handling of near vertical segments of curves where the slope is infinity. .

The parametric form of curve equation does not propose relation between x, y or z but considers them as independent variables each being a function of a scalar u. Figure 5.4 shows a typical 3D curve. The point on the curve P[x, y, z] can be represented as under

$$P[x, y, z] = [f_1(u), f_2(u), f_3(u)]$$

where $u \in [u_{min}, u_{max}]$

f_1, f_2, f_3 are functions in **u** which govern the x, y, z coordinates respectively.

Since the curve is conceived as a locus of point P, the above equation can be considered as a Vector valued parametric equation of the curve.

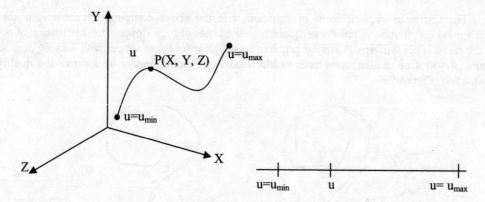

Figure 5.4 Curve in Cartesian and Parametric Space

The curve is conceptually mapped in the parametric domain in Figure 5.4. There exists one to one correspondence between the parametric **(u)** and the Cartesian **(R^3)** space. Thus,

$$\forall u \in [u_{min}, u_{max}] \exists P(x, y, z) \in R^3$$

The curve is a bounded parametric entity as **u_{min}, u_{max}** form the extreme ends of the curve. Without loss of generality, the parametric space can be normalized. The function could be denoted by **X (u), Y (u)** etc instead of f_1, f_2.

The generalized vector valued parametric equation for any curve is given by

$$P[u] = [X(u), Y(u), Z(u)], \forall u \in [0, 1] \qquad 5.7$$

The functions **X (u), Y (u)** etc will depend upon the type of curve to be represented. Few examples are included here to illustrate the use of vector valued parametric curve equations.

Straight Line

Vector valued parametric equations of a line passing through points **P_1 (1, 1)** and **P_2 (10, 8)** is given by

$$P[u] = P_1 + u (P_2 - P_1), \forall u \in [0, 1]$$
$$[X (u), Y (u)] = [X_1 Y_1] + u [(X_2 - X_1), (Y_2 - Y_1)]$$

The equation is

$$X (u) = x_1 + u (x_2 - x_1)$$
$$= 1 + 9u$$
$$Y (u) = y_1 + u (y_2 - y_1)$$
$$= 1 + 7u \qquad 5.8$$

Figure 5.5 show the Line and the variations of **X (u), Y (u)** in the parametric range. It can be seen that both **X (u) and Y (u)** are linear functions in **u** resulting in the curve representation as a straight Line **$P_1 P_2$**.

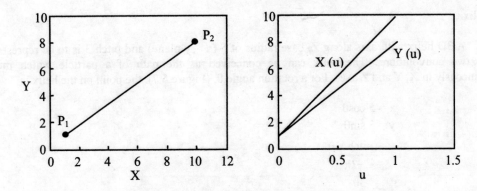

Figure 5.5 Line and the variations of the parametric functions

Circle

An arc of a circle with center at origin and radius of 5 is to be represented (Figure 5.6).

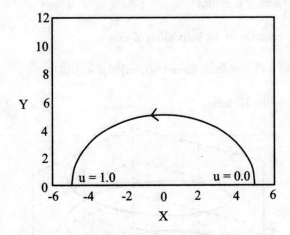

Figure 5.6 Arc of semi-circle

Using the polar form, the point on the circle is

$$x = r\cos\theta$$
$$y = r\sin\theta, \qquad\qquad \text{where } \theta \in [0,\pi] \text{ , } r = 5.0$$

Transforming to parametric space,

$$\theta = \pi u, \qquad\qquad \text{where } u \in [0,1]$$

The parametric equation of the arc of circle is

$$P[u] = [5\cos(\pi u),\ 5\sin(\pi u)] \ \forall \ u \in [0,1] \qquad\qquad \textbf{5.9}$$

3D Helix

A 3D helix with axis along Z, base radius of 5 (x – y plane) and pitch 3 is to be represented for its two convolutions. The helix can be conceived as the path of a particle which moves simultaneously in X, Y and Z axes. For a rotation angle θ, (Figure 5.7), the point on the helix is

$$x = 5 \cos\theta$$
$$y = 5 \sin\theta$$
$$z = \frac{pitch \times \theta}{2\pi}$$
$$= \frac{3\theta}{2\pi}$$

Transforming θ to parametric space for two convolutions of the helix

$$\theta = u\,\theta_{max}$$
$$= 4\pi u, \forall\, u \in [0,1]$$

Vector valued parametric equation of the helix along Z-axis is

$$P[u] = [5 \cos(4\pi u),\ 5 \sin(4\pi u),\ 6u]\ \forall\, u \in [0,1] \qquad\qquad \textbf{5.10}$$

Figure 5.7 shows the plot of the 3D helix.

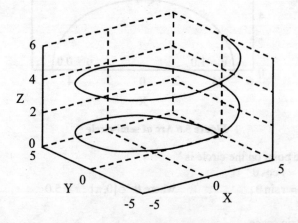

Figure 5.7 3D Helix

From the above examples it can be seen that a proper form of **X (u), Y (u), Z (u)** can be chosen to represent variety of planar (2D) and space (3D) curves.

5.4.4 Properties of Parametric Equations

Vector valued parametric curve equations possess some important properties which are advantageous in computer aided geometric design (CAGD).

- Since the curve equation is represented in parametric domain (u) and not the Cartesian one, the curve is axis independent. This is a very important property for curve synthesis.
- A single parameter **u** can be used to compute X, Y, Z coordinates. This is quite efficient in computational tasks
- Parameter **u** traces curve along its length rather than along the Cartesian axes. Points along the curve can, thus, be efficiently chosen to control the arc length.
- Curve segmentation and (re)parameterization is easy to handle segments of the curve.
- The curve can be differentiated in the parametric domain to compute property like tangent vector.

$$\frac{dp}{du} = P'(u) = \left[\frac{dX(u)}{du}, \frac{dY(u)}{du}, \frac{dZ(u)}{du} \right]$$

$$= [X'(u), Y'(u), Z'(u)]$$

where 'denotes differentiation. **P' (u)** is the Tangent Vector, to the curve at u. Cartesian slope can be obtained from the tangent vector. For example for a planar curve

$$P'(u) = [X'(u), Y'(u)]$$

$$\text{Slope } m = \frac{dy}{dx} = \frac{dy}{du} \bigg/ \frac{dx}{du} = Y'(u) / X'(u) \qquad \textbf{5.11}$$

By substituting appropriate values **[X' (u), Y' (u)]** in the Tangent Vector conditions of infinity slope can be easily handled.

- Due to the vector valued form, both Geometric and projection transformations discussed in chapter 3 and 4 can be easily applied to the curve.

In this book, vector valued parametric form of representation for curves and surfaces will be used.

5.5 CURVE DESIGN FROM POINTS.

Quite often it is desired to obtain the equation of the curve from points which are known apriori. The point data may be known by the observation of the physical phenomena (rainfall, sales data etc), by conducting an experiment to gather data or by inspecting a physical part (model) by coordinate measuring machine (CMM)/ optical scanner. To obtain curve equation from such data, two techniques are broadly followed viz. *Interpolation and Approximation*. Figure 5.8 shows the two typical situations.

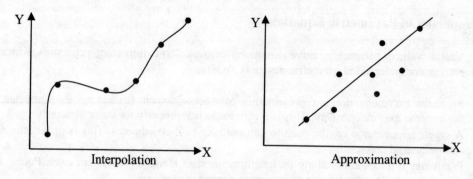

Figure 5.8 Curve Interpolation and Approximation

Interpolation aims at the design of the curve which passes through the data points. Shape of the curve in between the data points is dictated by the degree of the curve used for interpolation. Interpolation technique is used when the chosen data points are accurate and important. The curve must pass through them. Approximation, on the other hand, focuses on obtaining the curve equation from data points which are often not so accurate but show a general trend of variation of the data. The curve does not necessarily pass through the data points but gives a least square approximation of the trend. Such class of problems are collectively known as *Regression*. Several texts are available which discuss in details, the techniques for regression analysis. While these are used widely for several curve fitting applications, they are not suitable for CAGD applications where accuracy and shape control are important.

In what follows, two widely used Interpolation based techniques will be discussed

5.5.1 Lagrange Interpolation Technique

Lagrange's Interpolation method is a simple and elegant way of finding a unique interpolating polynomial **P(x)** of degree $\leq$ **(n-1)** which passes through **n** data points $[(x_1y_1), (x_2y_2) \ldots.(x_ny_n)]$. Once the curve equation is known, it can be used to compute value of function at any intermediate point.

Mathematical basis of the Lagrange interpolating polynomial **P(x)** is discussed below.
Let

$$P(x) = \sum_{j=1}^{n} P_j(x)$$

where

$$P_j(x) = y_j \prod_{\substack{k=1 \\ k \neq j}}^{n} \frac{(x - x_k)}{(x_j - x_k)} \qquad \textbf{5.12}$$

writing explicitly,

$$P(x) = \frac{(x - x_2)(x - x_3)\ldots\ldots(x - x_n)}{(x_1 - x_2)(x_1 - x_3)\ldots\ldots(x_1 - x_n)} y_1 + \cdots + \frac{(x - x_1)(x - x_2)\ldots\ldots(x - x_{n-1})}{(x_n - x_1)(x_n - x_2)\ldots\ldots(x_n - x_{n-1})} y_n \qquad \textbf{5.13}$$

This formula was first published by Waring in 1779, rediscovered by Euler in 1783 and published by Lagrange in 1795. It is sometimes referred as Waring - Lagrange interpolation formula.

While constructing the interpolating polynomial from a set of known data points, one can choose the degree of the interpolating curve. There is a tradeoff between a curve having a better fit and one having a smooth well behaved fitting function. If larger number of data points are used in the interpolation, the degree of resulting polynomial will be higher. This will result in greater oscillations of the curve between the data points. A high degree polynomial may, therefore, be a poor predictor of the function between the points, though the accuracy at the data points will be perfect.

Figure 5.9 shows typical Lagrange polynomials for various data points. Note the curve oscillations for larger number of points.

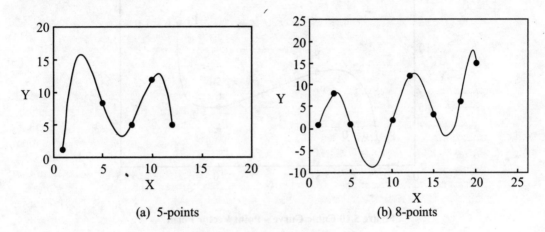

(a) 5-points   (b) 8-points

Figure 5.9 Lagrange Polynomial

5.5.2 Parametric Curve Equation from Points

Similar to the Lagrange Interpolation for the points in Cartesian space, interpolating polynomials can be obtained in parametric space to obtain curve equations from points.

In vector valued parametric form, a curve is represented by

$$P[u] = [X(u), Y(u), Z(u)], \ \forall \ u \in [0,1]$$

The functions **X(u), Y(u)** etc can be obtained from known data points say **P** [**P₁,P₂...Pₙ**]. This form of curve equation is termed as the point vector polynomial form of parametric curve. A typical example is included here to illustrate.

A vector valued parametric equation of a cubic planar curve is to be obtained from data points **P₀(x₀, y₀), P₁(x₁, y₁), P₂(x₂, y₂)** and **P₃(x₃, y₃)**. The curve equation can be written as

$$P[u] = [X(u), Y(u),], \ \forall \ u \in [0,1]$$

Each function can be of the form of a cubic polynomial in **u**. Thus,

$$P(u) = \sum_{i=0}^{3} a_i u^i$$
$$= a_0 + a_1 u + a_2 u^2 + a_3 u^3$$

5.14

Where a's are vector valued coefficients of the polynomial such as $a_0 = [a_{0x} \ a_{0y}]$ etc.

Figure 5.10 shows the conceptual curve in Cartesian and parametric spaces. The curve passes through points P_0, P_1, P_2, P_3 known apriori. The curve can be considered bounded and the extreme data points P_0 and P_3 lie at the two ends of parametric domain u =0, and u=1.0. It is presumed that values of **u** are known apriori at points P_1 and P_2

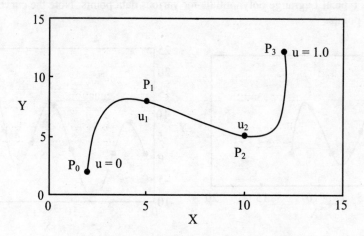

Figure 5.10 Cubic Curve – Point Vector Form

Writing out the equations in matrix form,

$$P(u) = \begin{bmatrix} 1 & u & u^2 & u^3 \end{bmatrix} \begin{bmatrix} a_0 \\ a_1 \\ a_2 \\ a_3 \end{bmatrix}$$

Substituting the boundary conditions,

$$\begin{bmatrix} P_0 \\ P_1 \\ P_2 \\ P_3 \end{bmatrix} = \begin{bmatrix} 1 & 0 & 0 & 0 \\ 1 & u_1 & u_1^2 & u_2^3 \\ 1 & u_2 & u_2^2 & u_2^3 \\ 1 & 1 & 1 & 1 \end{bmatrix} \begin{bmatrix} a_0 \\ a_1 \\ a_2 \\ a_3 \end{bmatrix} \qquad 5.15$$

For known values of P_0, P_1, P_2, P_3 and corresponding **u** values, the coefficients $[a_0 \ a_1 \ a_2 \ a_3]$ can be computed. Using these, the curve equation (Equation 5.15) can be obtained and used for interpolation.

The technique can be applied to both 2D and 3D curves with varying number of data points. As seen the degree of the curve is **n** (cubic in this case) when **n+1** are the number of data points. As before, higher degree curves introduce more oscillations in the curve which are undesirable. As a result, cubic curves are more widely used for design of 3D space curves.

5.5.3 Hermite curves

A parametric cubic curve is represented in terms of a cubic polynomial function as under

$$P(u) = \sum_{i=0}^{3} a_i u^i$$

$$= \begin{bmatrix} 1 & u & u^2 u^3 \end{bmatrix} \begin{bmatrix} a_0 \\ a_1 \\ a_2 \\ a_3 \end{bmatrix}$$

5.16

The vector coefficients $[a_0 \ a_1 \ a_2 \ a_3]$ need to be evaluated for specific boundary conditions. A cubic curve needs four boundary conditions to obtain the curve equation.

In comparison to the Point vector form outlined in the previous section, Hermite curves use different boundary conditions. Figure 5.11 shows a typical Hermite curve and the associated boundary conditions.

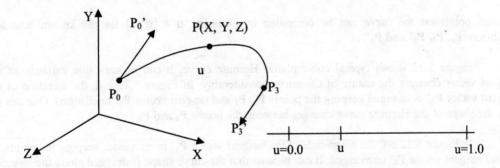

Figure 5.11 Hermite Cubic Curve – Boundary Conditions

The curve is defined in terms of its end points P_0, P_3 and the tangent vectors to the curve at end points P_0' and P_3'. Conceptually the curve can be conceived to be the locus of a point which starts from P_0 along the direction specified by tangent vector P_0'. The curve ends at P_3 along the direction specified by P_3'. Without loss of generality, the curve can be parameterized in terms of $u \in [0, 1]$ from P_0 to P_3. Shape of the curve will be invariant with the direction of parameterization.

The boundary conditions for curve design are,
$$u = 0; P = P_0, \text{tangent vector} = P_0'$$

$$u = 1; P = P_3, \text{tangent vector} = P_3'$$

Differentiating the curve Equation 5.16, the tangent vector is

$$P'(u) = \begin{bmatrix} 0 & 1 & 2u & 3u^2 \end{bmatrix} \begin{bmatrix} a_0 \\ a_1 \\ a_2 \\ a_3 \end{bmatrix}$$

Substituting the boundary conditions,

$$\begin{bmatrix} P_0 \\ P_3 \\ P_0{}' \\ P_3{}' \end{bmatrix} = \begin{bmatrix} 1 & 0 & 0 & 0 \\ 1 & 1 & 1 & 1 \\ 0 & 1 & 0 & 0 \\ 0 & 1 & 2 & 3 \end{bmatrix} \begin{bmatrix} a_0 \\ a_1 \\ a_2 \\ a_3 \end{bmatrix}$$

5.17

Knowing P_0, $P_0{}'$, P_3, $P_3{}'$, Equation 5.17 can be solved to obtain values of vector coefficients $[a_0\ a_1\ a_2\ a_3]$. Substituting these into Equation 5.17 the equation for parametric cubic Hermite curve can be written as

$$P(u) = a_0 + a_1 u + a_2 u^2 + a_3 u^3$$

where

$$a_0 = P_0 , a_1 = P_0{}'$$
$$a_2 = -3\ P_0 - 2\ P_0{}' + 3\ P_3 - P_3{}'$$
$$a_3 = 2\ P_0 + P_0{}' - 2\ P_3 + P_3{}'$$

5.18

Various points on the curve can be computed by choosing $u \in [0, 1]$ for the known boundary conditions P_0, P_3, $P_0{}'$ **and** $P_3{}'$.

Figure 5.12 shows typical cubic planar Hermite curve. It can be seen that variation of the tangent vector changes the nature of the curve considerably. In Figure 5.12 a-d, the direction of the tangent vector $P_3{}'$ is changed keeping the points P_0, P_3 and tangent vector $P_0{}'$ unchanged. One can see how the shape of the Hermite curve changes between the points P_0 and P_3.

In Figure 5.12 e-f, the magnitude of the tangent vector $P_0{}'$ is increased, keeping the points P_0, P_3 and tangent vector $P_3{}'$ unchanged. It can be seen that the curve shape is dragged along the direction of $P_0{}'$. The curve is flexible and smooth between the chosen end points P_0 **and** P_3.

In conclusion it can be stated that increasing the magnitude of tangent vector drags the curve along its direction while changing direction of tangent vector causes changes in the curvature of the curve, introducing the concave/convex regions.

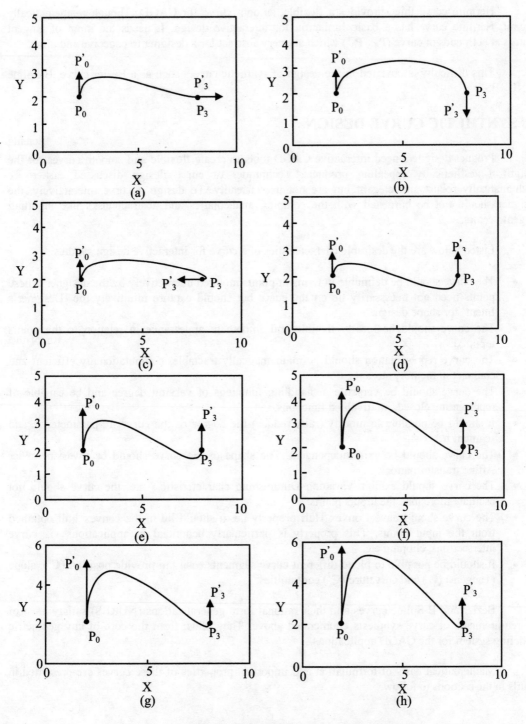

Figure 5.12 Hermite Curves – Shape Control

Hermite curve thus, provides a flexible, smooth curve for CAGD. Though mathematically elegant, Hermite curve has a basic limitation for interactive design. It needs the input of tangent vectors at both ends of curve (P_0', P_3') which are very abstract for a designer to conceive and input.

This difficulty is obviated in the design of synthetic curves such as a Bezier curve, B-Spline or NURBS.

5.6 SYNTHETIC CURVE DESIGN

Product designers need interactive CAGD tools to create flexible and smooth curves for the design of aesthetically appealing products. Techniques of curve design discussed earlier are mathematically robust and elegant but are not user friendly. To design a curve interactively, the designer should not be burdened with the complex mathematics and abstract tasks like inputting tangent vectors.

Listed below are the desirable characteristics of a curve for interactive design synthesis.

- The curve should be definable in terms of points inputted interactively by the designer. These points need not necessarily lie on the curve but should capture intuitively the 'Designer's Intent' for shape design.
- The curve should be locally flexible and smooth in appearance in relation to the points inputted.
- The curve representation should be mathematically tractable, computationally efficient and compact in memory storage.
- The curve should be versatile in handling functions of varying degree and be capable of representing closed, multivalued functions.
- It should be possible to modify (raise/reduce) the degree of the curve, re-parameterize and segment it.
- The curve should be axis independent. The shape of the curve should be invariant under Affine transformations.
- The curve should exhibit Variation Diminishing characteristic – i.e. the curve should not oscillate more than the inputs points.
- The curve should have Convex Hull property i.e. it should lie in the convex hull obtained from the input points. This property is particularly beneficial in applications viz curve intersections, clipping etc.
- It should be possible to blend different curve segments easily to provide position (C^0), slope (Tangent) (C^1) and curvature (C^2) continuities.

Bezier and B-Spline curves and their rational form generalizations (NURBS) satisfy most of the requirements of curve synthesis enumerated above. They, thus, form the core of any geometric modeling system for the CAGD applications.

Mathematical basis of formulation and important properties of these curves are presented in details in the sections to follow.

5.7 BEZIER CURVES

Bezier curve is a very important synthetic curve used in CAGD, art and animation. It was invented by a French Engineer P. Bezier in early 70's for the design of complex surfaces of Renault cars. Lot of research has been carried out on Bezier curves and surfaces by researchers from mathematics, computer science and CAD/CAM disciplines.

In what follows, mathematical basis for the synthesis of Bezier curve and its fundamental properties are presented.

5.7.1 Bezier Curve Equation

Bezier curve is a flexible curve which is designed to suit the needs of ab initio shape synthesis. No prior data is required by the designer about the points on the curve or its properties like tangent vector.

Bezier curve is designed by specifying points termed as control points, all of which do not lie on the curve but govern its overall shape. The curve generally follows the shape specified by the control points (control polygon) and is smooth (fluid). By dragging the control points, the designer can easily manipulate the shape of the curve. It thus, provides a very user friendly interactive design-synthesis environment.

Mathematically, the equation of the Bezier curve is represented as under.
Given a set of **(n+1)** control points [P_0, P_1, P_2............ P_n], the vector valued parametric equation of the Bezier curve of degree **n** is given by

$$P(u) = \sum_{i=0}^{n} B_{i,n}(u) \, P_i \qquad\qquad 5.19$$

where $B_{i,n}$ are the Berstein Basis functions given by

$$B_{i,n}(u) = \binom{n}{i} u^i (1-u)^{n-i}$$

$\binom{n}{i}$ is a Binomial Coefficient ^nC_i,

Thus, $\qquad\qquad B_{i,n}(u) = \dfrac{n!}{i!(n-i)!} u^i (1-u)^{n-i}, \forall u \in [0,1]$

Writing out,
$$P(u) = B_{0,n}(u)\, P_0 + B_{1,n}(u)\, P_1 + \ldots\ldots\ldots + B_{n,n}(u)\, P_n \qquad\qquad 5.20$$

Equation 5.20 shows that a position vector **P (u)** on the Bezier curve is a weighted average of the control points [P_0, P_1 ... P_n]. The weightage functions are the Basis functions [$B_{0,n}$, $B_{1,n}$,......... $B_{n,n}$] which are the polynomial functions of **u** with degree **n**.

The Basis functions termed as Blending functions, give some interesting and important properties to the Bezier curve. These are explained by taking an example of a cubic Bezier curve.

5.7.2 Cubic Bezier Curve

A cubic Bezier curve (degree n = 3) is defined in terms of 4 control points. Figure 5.13-shows the control vertices [P_0, P_1, P_2, P_3] and the resulting Bezier curve.

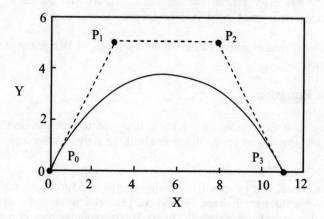

Figure 5.13 Bezier Curve and Control Points

Vector valued parametric equation of the curve is given by

$$P(u) = \sum_{i=0}^{3} B_{i,3}(u)\, P_i$$
$$= B_{0,3}(u)\, P_0 + B_{1,3}(u)\, P_1 + B_{2,3}(u)\, P_2 + B_{3,3}(u)\, P_3$$

The Basis function $B_{i,3}(u)$ are termed as Blending functions as they 'blend' the control points to govern a position vector (point) on the curve. For a cubic curve, the blending function is as under

$$B_{i,n}(u) = {}^3C_i\, u^i\, (1-u)^{3-i}$$

Thus,

$$B_{0,3}(u) = B_{i,n}(u) = \frac{3!}{0!\,3!} u^0 \, (1-u)^3$$
$$= (1-u)^3$$

In a similar way,

$$B_{1,3}(u) = 3u(1-u)^2$$
$$B_{2,3}(u) = 3u^2\,(1-u)$$
$$B_{3,3}(u) = u^3$$

5.21

The curve is parameterized from P_0 to P_3 such that $u \in [0, 1]$. Knowing the control points [P_0, P_1, P_2, P3], various points on the curve can be computed using Equation 5.20 and Equation 5.21. Blending functions $B_{i,n}(u)$ are cubic in nature. Figure 5.14 shows the variation of blending functions [$B_{0,3}$, $B_{1,3}$, $B_{2,3}$, $B_{3,3}$] in the parametric range [0,1]. It can be seen that all blending are non zero in the entire parametric range. At the start (u=0) and end (u=1) of the curve, the respective blending functions $B_{0,3}$ and $B_{1,3}$ are 1, while all the other function are zero. This makes the curve to pass through P_0 and P_3.

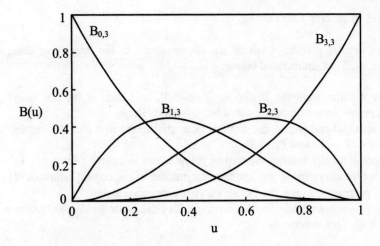

Figure 5.14 Variation of Cubic Bezier Blending Functions

Figure 5.15 shows variety of cubic Bezier curves which are obtained from different control points. Flexibility and smoothness (Fluidity) of the Bezier curve can be seen from these examples.

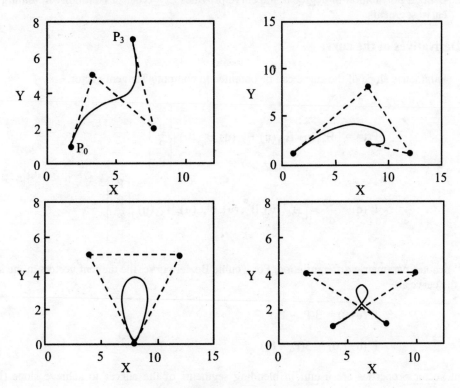

Figure 5.15 Cubic Bezier Curves

5.7.3 Properties of Bezier Curve

Bezier curve has several properties which are advantageous in the interactive design environment. Important among them are enumerated below.

- The curve starts from the first control point P_0 and ends at the last one P_n. Intermediate control points do not lie on the curve but govern its shape.
- At the start and end points, the curve is tangential to the first and last segments of the control polygon viz $P_0 - P_1$ and $P_{n-1} - P_n$.
- The curve generally follows the control polygon and is smooth (fluid).
- Degree of the curve **(n)** is one less than the number of the control points **(n+1)**.
- Being a parametric curve, the Bezier curve is axis independent.
- The curve lies completely in the convex hull of the control polygon shown as dotted line in Figure 5.13. This is because

$$\sum_{i=0}^{n} B_{i,n}(u) = 1$$

- The curve exhibits variation diminishing characteristic i.e. the curve does not oscillate more than the control points.
- Being a polynomial of degree n, the curve provides C^{n-1} order of continuity in joining various curve segments.

5.7.4 Derivatives of the curve

Parametric slope of the curve can be obtained to compute Tangent vector.

$$P(u) = \left[B_{0,3}(u), B_{1,3}(u), B_{2,3}(u), B_{3,3}(u) \right] \begin{bmatrix} P_0 \\ P_1 \\ P_2 \\ P_3 \end{bmatrix}$$

$$P'(u) = \frac{dp}{du} = \left[B'_{0,3}(u), B'_{1,3}(u), B'_{2,3}(u), B'_{3,3}(u) \right] \begin{bmatrix} P_0 \\ P_1 \\ P_2 \\ P_3 \end{bmatrix}$$

where ' denotes parametric differentiation. For a cubic Bezier curve, the tangent vectors at the start and end of the curve are

$$P'(u=0) = 3(P_1 - P_0)$$

$$P'(u=1.0) = 3(P_3 - P_2) \hspace{4cm} \textbf{5.22}$$

Tangent vector properties are useful in blending segments of the curves to achieve slope (tangent) continuity **C'**.

5.7.5 Subdivision of the Curve

Many a times, it is required to subdivide a curve into segments to handle tasks like intersection, clipping etc. De Casteljau has given an interesting and efficient algorithm for subdividing a Bezier curve from its control polygon. The algorithm is explained here for curve bisection.

De Casteljau's Subdivision Algorithm

It is desired to bisect the cubic Bezier curve to obtain its parametric midpoint (u = 0.5). Figure 5.16 shows the recursive procedure of the algorithm.

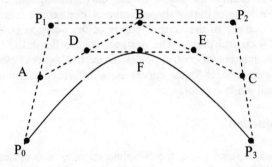

Figure 5.16 De Casteljau's Recursive Subdivision Procedure

The computing steps are as under
- Obtain n-spans of the control polygon viz $P_0 - P_1$, $P_1 - P_2$, and $P_2 - P_3$
- Compute midpoint of each segment (span) viz **A, B, C**.
- Repeat recursively till further reduction is not possible
- The final point obtained (**F**) is the desired midpoint of the curve (u=0.5).

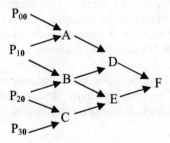

Figure 5.17 De Casteljau's Algorithm – Scheme of Computation

Figure 5.17 shows the scheme of computing for subdividing the cubic Bezier curve in Figure 5.16 It is interesting to see that the two curve segments P_0F and FP_3 are joined at **F** with tangent (**C'**) continuity and point sets [P_0 A D F] and [F, E, C, P_3] from the respective control polygons for the two curves segments.

Typical examples on Bezier curve design and its properties are included at the end of this chapter.

5.8 B – SPLINE CURVES

Bezier curve is a flexible and smooth curve for interactive design synthesis. Though used in many CAGD applications, the Bezier curve has two major limitations which restrict its flexibility and interactivity.

The Bezier curve uses Berstein basis (blending) functions to obtain point on the curve in terms of the control points. The basis functions are non zero in the entire parametric range [0,1]. (Figure 5.14) Since the point on the curve is the weighted average of the control polygon vertices (points), the curve does not provide *Local Control* over the shape of the curve. Thus, the movement of a single control point will change the shape of the curve throughout its length. This could be particularly annoying to the designer during shape synthesis. Further the degree **(n)** of a Bezier curve is always one less than the number of the control points of the polygon **(n+1)**. For example a cubic Bezier curve would need 4 control points. The designer has no flexibility to alter the degree of the curve without changing the number of control points. These two factors restrict the flexibility offered by a Bezier curve during synthesis. B-Spline curves use a different Basis function which provides the required flexibility and local shape control.

5.8.1 B-Spline Basis Function

From a mathematical stand point, the B-Spline curve is also obtained from a set of control points. Thus a point on the B-Spline curve is obtained as the weighted average of the control points; the weightage being provided by the B-Spline Basis (blending) function. The key difference lies in the formulation of the B-Spline Basis function which provides the desired characteristics of flexibility and local shape control.

For B-Spline curve design, there is one basis function associated with each control point. The basis function is Non Zero over a specific range of the parametric space. Due to this non global nature of the basis function, the influence of the associated control point in governing the shape of the curve is felt over the limited parametric (curve) domain. This property provides the *Local Control* over the shape of the curve. Further the nature of the blending function can be turned to change the degree of the curve without changing the number of control points. In essence, B-Spline basis function can be considered to be a superset of the Berstein basis function.

Recursive definition of the B-Spline basis function suitable for numerical computation was independently discovered by Cox and De Boor. Gorden and Rosenfield further applied the basis function to curve design. In this book, the recursive definition of B-Spline Basis function given by Cox and De Boor will be followed.

5.8.2 B-Spline curve Equation

The B-Spline curve is defined in terms of (n+1) control points $[P_1, P_2, P_3 \dots P_{n+1}]$ in order. The vector valued parametric equation of the curve is given by

$$P(u) = \sum_{i=1}^{n+1} B_{i,k}(u)\, P_i \qquad\qquad \textbf{5.23}$$

where $B_{i,k}(u)$ are the B-Spline basis functions, P_i represents the $(n+1)$ vertices of the control polygon. The parameter u is bounded $u \in [u_{min}, u_{max}]$. k is the order of B-Spline basis function (degree k-1) such that $2 \le k \le n+1$. $B_{i,k}(u)$ are the normalized B-Spline basis functions which are non zero over a specific range of **u** .

Using Cox -De Boor recursive formula

$$B_{i,1}(u) = 1 \qquad \text{if } t_i \le u < t_i + 1$$
$$= 0 \qquad \text{otherwise}$$

$$B_{i,k}(u) = \frac{(u - t_i) B_{i,k-1}(u)}{(t_{i+k-1} - t_i)} + \frac{(t_{i+k} - u) B_{i+1,k-1}(u)}{(t_{i+k} - t_{i+1})} \qquad \text{5.24}$$

In order to get $B_{i,k}(\mathbf{u})$ non zero in a specific range, the domain **u** is split into segments using a knot vector **T**. The elements of the knot vector **T** are the values of t_i satisfying the relation $t_i \le t_{i+1}$. The parameter **u** varies from $\mathbf{u_{min}}$ to $\mathbf{u_{max}}$ along the curve. The recursive relation assumes the convention 0/0=0.

$\mathbf{B_{i,k}}$ is the polynomial basis function of order **k** (degree k-1) which satisfies the following conditions,

- The basis function $\mathbf{B_{i,k}(u)}$ is a polynomial of degree (k-1) on each interval $\mathbf{t_i \le u < t_{i+1}}$.
- The curve equation **P(u)** and its derivatives of the order 1,2,........k-2 are all continuous over the entire parametric range of the curve.
- For k=2, the curve degenerates to the control polygon while for k = (n+1), the curve becomes a Bezier curve providing global basis functions.

To design a B-Spline curve, three broad steps need to be followed.

- Choose a set of control points $[\mathbf{P_1, P_2, ... P_{n+1}}]$ in order which form the general shape of the curve.
- Choose the basis order $\mathbf{k}$ $(\mathbf{2 \le k \le n+1})$ which decides the degree of the curve.
- Choose the knot vector **T** $[\mathbf{t_1, t_2, t_{3,.........}}]$ by segmenting the parameter domain **u** into spans.

Knot vector is a set of monotonically increasing real numbers $[\mathbf{t_1, t_2, t_{3,.........}}]$ such that $\mathbf{t_i \le t_{i+1}}$. Various strategies to choose a knot vector are discussed below

5.8.3 B-Spline Basis Function and Knot vector

Choice of a knot vector has a significant influence on the B-Spline basis function $\mathbf{B_{i,k}(u)}$ and consequently on the shape of the curve. Three strategies exist for the choice of knot vectors viz *uniform (Periodic), Open Uniform and Non Uniform*.

Uniform knot vector

For a uniform knot vector, the knot values t_i can be evenly spaced. For example
$$T = [0\ 1\ 2\ 3\ 4]$$

Alternatively the knot values can start from 0 and be incremented in normalized space upto 1.0. For example
$$T = \left[0\ \frac{1}{4}\ \frac{2}{4}\ \frac{3}{4}\ \frac{4}{4}\right]$$
$$T = [0\ 0.25\ 0.5\ 0.75\ 1.0]$$

To understand the variations of the B-Spline basis function with knot vector, a specific example is included.

Let the control polygon be defined by the control points $[P_1, P_2, P_3, P_4]$ and the curve order k=3 (degree n=2). Number of knots needed to design the curve will be n+k+1= 7. Using uniform knot strategy, the knot vector will be
$$T = [0\ 1\ 2\ 3\ 4\ 5\ 6]$$

Using Cox -De Boor recursive relation (Equation 5.22), the scheme of computation is shown in Figure 5.18

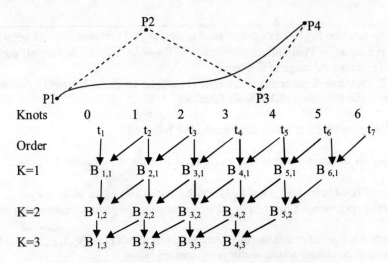

Figure 5.18 Cox De Boor recursive computation scheme

Equation 5.22 shows that the basis function $B_{i,k}(u)$ is derived from $B_{i,k-1}(u)$ and $B_{i+1,k-1}(u)$. Computations are shown here for a typical blending function say $B_{1,3}(u)$.

Step 1

Using Cox -De Boor relation, blending function $B_{1,1}$, $B_{2,1}$, $B_{3,1}$, $B_{4,1}$...etc are 1 in the respective knot span range [0,1), [1,2), [2,3), [3,4) etc. Figure 5.19 shows the variation of these blending functions

Step 2

Using Equation 5.22 and the blending function $B_{i,1}(u)$ computed earlier, $B_{1,2}(u)$ can be computed. For example

$$B_{1,2}(u) = \frac{(u-t_1)}{(t_2-t_1)}B_{1,1}(u) + \frac{(t_3-u)}{(t_3-t_2)}B_{1,1}(u)$$

Uniform knot vector T = [0 1 2 3 4 5 6], the basis function is

$$\begin{aligned}B_{1,2} &= u & 0 \le u < 1 \\ &= (2-u) & 1 \le u < 2\end{aligned}$$

Blending functions $B_{2,2}$, $B_{3,2}$,... will be translates of $B_{1,2} = u$ along the parametric space. Figure 5.19 shows the various blending functions.

Step 3

Proceeding further on similar lines, the basis function $B_{1,3}(u)$ will be

$$\begin{aligned}B_{1,3} &= 0.5u^2 & 0 \le u < 1 \\ &= 0.5(-3 + 6u - 2u^2) & 1 \le u < 2 \\ &= 0.5(3-u)^2 & 2 \le u < 3\end{aligned}$$

As earlier, other blending functions $B_{2,3}$, $B_{3,3}$,... will be translates of $B_{1,3}(u)$. Figure 5.19 shows the plot of the blending functions.

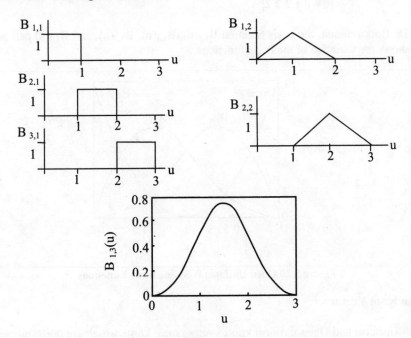

Figure 5.19 Uniform B-Spline Basis Functions

Figure 5.19 shows that a typical blending function $B_{1,3}$ is spread over 3 segment spans of the tangent vector $T = [0\ 1\ 2\ 3]$ viz 0-1, 1-2 and 2-3. The basis function is a curve of degree 2 (k=3) whose segments are joined smoothly at the knots viz u=1 and u=2. An important property of the basis function is seen that the functions are non zero in a specific parameter range decided by the knot values.

Open Uniform Knot Vector

An open uniform knot vector is recommended to design curves that behave quite similar to Bezier curves. The knot values are replicated at the ends of parameter span equal to the order **k** of the basis function. This essentially ensures that the curve passes through the start and end points of the control polygon like the Bezier curve.

As earlier, for control points $[P_1, P_2, P_3, P_4]$ and basis order **k=3** (degree 2), the number of knots = n+k+1 = 7.

An open uniform knot vector is formally given by
$$t_i = 0 \qquad 1 \leq i \leq k$$
$$= i - k \qquad k+1 \leq i \leq n+1$$
$$= n-k+2 \qquad n+2 \leq i \leq n+k+1$$

Thus, the knot vector will be
$$T = [0\ 0\ 0\ 1\ 2\ 2\ 2]$$

Using Cox -De Boor relation, the basis function $B_{1,3}(u)$, $B_{2,3}(u)$, $B_{3,3}(u)$, and $B_{4,3}(u)$ can be computed. Figure 5.20 shows the variation of the basis functions.

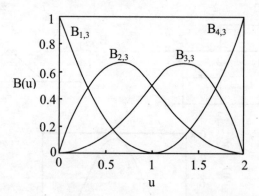

Figure 5.20 Open Uniform B-Spline Basis Functions

Non Uniform Knot Vector

Both Uniform and Open uniform knots vectors have knots which are uniformly spaced along the parameter space. In contrast, the non uniform knot vector can have either unequally spaced and/ or multiple (replicated) knots. Further they could be Periodic or Open.

Typically the non uniform knot vectors could be of the form.

$$T = [0\ 0\ 0\ 1\ 1\ 2\ 2\ 2]$$
$$T = [0\ 1\ 2\ 2\ 3\ 4]$$
$$T = [0\ 1\ 1.8\ \ 2.2\ 3\ 4]$$

Uniform knot vectors create symmetric blending function (Figure5.19, 5.20) while non uniform knots introduce asymmetries in the basis function both in terms of range of u (knot spans) as well as the shape of function. Multiple knots introduce cusps in the basis functions. Figures 5.21 show some typical variations in the basis functions due to non uniform and multiple knots for **n+1 = 7, K = 3** and knot vector **T = [0 0 0 0.3 0.5 0.5 0.6 1 1 1]**

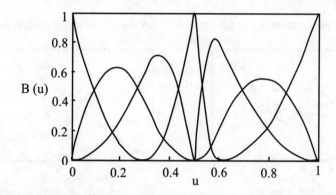

Figure 5.21 Non Uniform B-Spline Basis Function

5.8.4 Properties of B- Spline Curve

B- Spline Curve retains all the important properties of Bezier curve enumerated in sec 5.7. In addition, it provides a local control over the shape of the curve. In summary, important properties of the B-Spline curve can be enumerated as under

- The curve is definable in terms of control points. It generally follows the shape of the control polygon.
- The degree of the curve is not tightly linked with the number of control points. User can create family of curves from the same control polygon by varying the order **K** (degree **K-1**) between **K = 2** to **K = n+1**.
- The curve exhibits local control over the shape as the blending (basis) functions are non zero in a defined parametric range.
- The curve exhibits variation diminishing property i.e. it does not oscillate about any straight line more often than the control polygon.
- The curve lies within the convex hull of the control polygon. In fact the convex hull property of the B-Spline curve is stronger than the Bezier. The B-Spline curve lies in the convex hull formed by **K** neighboring vertices.
- The curve is axis independent.
- The shape of the curve is invariant under the affine transformations.
- The curve is smooth, fluid and offers properties of segmentation and reparameterization.

5.8.5 Controlling Shape of B- Spline Curve

A designer can interactively change the shape of curve using any of the flexible handles outlined below.

- Changing the control polygon in terms of number of control points and their positions
- Changing order of the basis function (k)
- Changing the knot vector – Periodic uniform, Open uniform, Non uniform.
- Introducing multiple (repetitive) knot values in the knot vector
- Introducing multiple (coincident) control points in the polygon

Figure 5.22 shows typical variations in the shape of B-Spline curves under different design conditions.

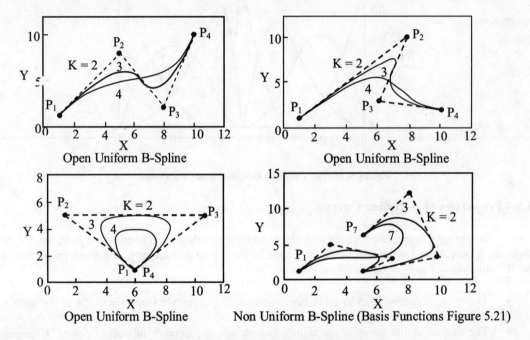

Figure 5.22 Shape Control of B-Spline Curve

It can be seen that at **K = 2**, the B-Spline curve is co-incident with the control polygon itself (shown by dotted line). For **K = N+1**, the curve becomes the Bezier curve, which is the loosest curve fitting within the control polygon.

5.9 NURBS CURVE

B-spline curve provide several nice properties to a designer in terms of curve synthesis, flexibility and shape control as enumerated earlier. However the B-spline curves are not able to represent simple conic curves like circle, ellipse or hyperbola mainly because the basis functions are polynomial. This difficulty is solved by using rational form of the basis functions.

NURBS is an acronym for Non Uniform Rational B-Spline which provides a single precise mathematical form capable of representing the commonly used analytical curves such as lines, conic curves, freeform curves, used in computer graphics and CAGD. NURBS have been studied in depth by several researchers like Piegl, Tiller, Barnhill to name a few. Today NURBS curves have been included in the computer graphics standards such as IGES and STEP. They have also been incorporated in the commercial geometric modeling Kernel/systems.

Mathematical basis of NURBS curve and its important properties will be presented in the sections to follow.

5.9.1 Rational B-Spline Curve.

A rational B-Spline curve is quite similar in its mathematical form to the non rational one except its basis function. In essence, a rational B-spline curve is the projection of a non rational (polynomial) B- spline curve defined in 4D homogeneous coordinate system into the 3D cartesian physical space.

Let the control polygon be defined by a set of control points $P_1^h, P_2^h \ldots\ldots P_1^h, P_{n+1}^h$

$$P_1^h = [w_1 x_1, w_1 y_1, w_1 z_1, w_1]$$

where w_i is a non zero scalar associated with each point in the homogenous coordinate space.
Vector valued parametric equation of the curve is

$$P(u) = \sum_{i=1}^{n+1} B_{i,k}(u)\, P_i \qquad\qquad 5.25$$

$B_{i,k}(u)$ is the non rational B-Spline basis function discussed earlier.(Section **5.8**)

Projecting the 4D homogeneous coordinates into 3D Cartesian space, the equation of the rational B-Spline becomes

$$P(u) = \frac{\sum\limits_{i=1}^{n+1} B_{i,k}(u).w_i.P_i}{\sum\limits_{i=1}^{n+1} w_i.B_{i,k}(u)}$$

$$= \sum_{i=1}^{n+1} R_{i,k}(u) P_i$$

Where $R_{i,k}(u)$ is the rational basis function given by

$$R_{i,k}(u) = \frac{w_i\, B_{i,k}(u)}{\sum\limits_{i=1}^{n+1} w_i.B_{i,k}(u)} \qquad\qquad 5.27$$

Associated with each vertex P_i, there is a blending function $\mathbf{R_{i,k}(u)}$ (≥ 0) and weightage $\mathbf{w_i}$. The weightage vector $\mathbf{w}$ [$\mathbf{w_1}$, $\mathbf{w_2}$......$\mathbf{w_n}$] provides additional flexibility to the designer to create family of curves from the same control polygon.

The rational B-Spline curve retains all the properties of B-Spline curves discussed in section 5.8 and provides one additional property that the curve shape will be invariant under projective transformation. Rational B-Spline curves are thus, a true superset of B-Spline as well as Bezier curves.

5.9.2 Rational Bezier Curve

As stated earlier, rational curves have the capability to represent conic curves like circle. Properties of a rational Bezier curve to represent conic curves are illustrated here with an example.

Let a quadratic Bezier curve (degree 2) be used to represent the conic curves. The control polygon $\mathbf{P}$ [$\mathbf{P_1}$, $\mathbf{P_2}$, $\mathbf{P_3}$] is associated with weightage vector $\mathbf{w}$ [$\mathbf{w_1}$, $\mathbf{w_2}$, $\mathbf{w_3}$]. Without loss of generality, let $\mathbf{w}$, be [$\mathbf{1\ w_2\ 1}$]. Vector valued parametric equation of the rational curve will be

$$P(u) = \frac{\sum\limits_{i=1}^{3} B_i(u).w_i.P_i}{\sum\limits_{i=1}^{3} B_i(u)w_i} \qquad \qquad \textbf{5.28}$$

where the blending function are given by $B_1 = (1-u)^2$, $B_2 = 2u(1-u)$, $B_3 = u^2$

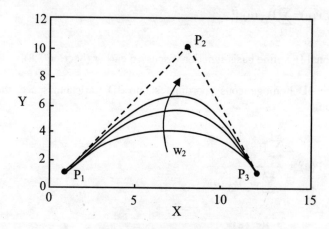

Figure 5.23 Rational Bezier Curve

Figure 5.23 Shows the conic sections represented by the rational Bezier curve. For different values of w_2, the following curve segments result between $\mathbf{P_1}$ and $\mathbf{P_2}$

$$w_2 = 0 \quad \text{- Straight line}$$
$$0 < w_2 < 1 \quad \text{- Elliptic arc}$$
$$w_2 = 1 \quad \text{- Parabolic arc}$$
$$w_2 > 1 \quad \text{- Hyperbolic arc}$$

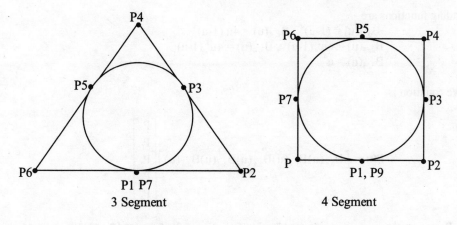

3 Segment   4 Segment

Figure 5.24 Full Circle Representation

Circle is a special case of an ellipse. By symmetry, a full circle can be formed by joining multiple segments. Figure 5.24 shows a typical full circle obtained from three quadratic rational B-Spline segments, each subtending an angle of 120° at the centre. Alternatively as shown, four curve segments can also be used, each subtending an angle of 90°. Appropriate control points $P[P_1, P_2,.....P_4]$ and the weightage vector $w[w_1, w_2....]$ can be chosen to create the three/four curve segments for designing the full circle.

5.10 EXAMPLES

1. A Bezier curve is designed from the control points $P_0[5, 0]$, $P_1[3, 8]$, $P_2[8, 7]$, $P_3[5, 2]$ and $P_4[12, 5]$.

   i) Derive vector valued parametric equation of the curve and compute typical points on it. Draw a neat sketch showing the curve and the control points.

   ii) Compute the tangent vectors at the start and end points of the curve. Verify if the curve satisfies convex hull property.

Solution:

i)   Vector valued parametric equation of the Bezier curve is given by

$$P(u) = \sum_{i=0}^{4} B_{i,4}(u)P_i, \forall u \in [0,1]$$

The basis function $B_{i,4}(u)$ are given by

$$B_{i,4}(u) = {}^4C_i \, u^i \, (1-u)^{4-i}$$

$$= \frac{4!}{i!(4-i)!} u^i (1-u)^{4-i}$$

where $i \in [0, 1, 2, 3, 4]$

The blending functions are

$$B_{0,4}(u) = (1-u)^4 \ , \ B_{1,4}(u) = 4u \, (1-u)^3$$
$$B_{2,4}(u) = 6u^2 \, (1-u)^2 \ , \ B_{3,4}(u) = 4u^3 \, (1-u)$$
$$B_{4,4}(u) = u^4$$

The curve equation is

$$P(u) = \begin{bmatrix} B_{0,4}(u) & B_{1,4}(u) & B_{2,4}(u) & B_{3,4}(u) & B_{4,4}(u) \end{bmatrix} \begin{bmatrix} P_0 \\ P_1 \\ P_2 \\ P_3 \\ P_4 \end{bmatrix} \qquad \textbf{5.29}$$

Points on the curve are computed using Equation 5.29 for $u \in [0, 1]$. Figure 5.25 shows the plot of the curve along with the control points.

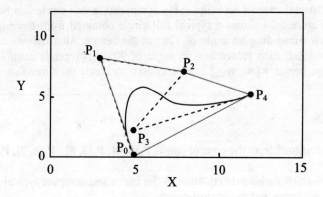

Figure 5.25

ii) Tangent vector at any point on the curve is obtained by differentiating the curve equation Equation(5.28) with **u**

$$P'(u) = \begin{bmatrix} B'_{0,4}(u) & B'_{1,4}(u) & B'_{2,4}(u) & B'_{3,4}(u) & B'_{4,4}(u) \end{bmatrix} \begin{bmatrix} P_0 \\ P_1 \\ P_2 \\ P_3 \end{bmatrix}$$

where, $B'_{i,4}(u) = \dfrac{d}{du}[B_{1,4}(u)]$

Differentiating and inputting boundary conditions, one gets

At u=0, $\qquad P'(u) = 4(P_1 - P_0)$
$\qquad\qquad\qquad\quad = (-8, 32)$
At u=1.0, $\qquad P'(u) = 4(P_4 - P_3)$
$\qquad\qquad\qquad\quad\ = (28, 12)$

Figure 5.25 shows that the curve is tangent to the First and Last control polygon spans at the start and end of the curve.

The convex hull of the control polygon is given by **[P₀ P₁ P₂ P₃]**. This is shown by the thin line in Figure 5.25. The curve completely lies in the convex hull, thus satisfying the property. This can also be verified by checking that $\sum_{i=0}^{4} B_{i,4}(u) = 1.0$.

2. A cubic Bezier curve is designed from the control vertices **P₀** [0, 0], **P₁**[5, 8] **P₂** [8, 5], **P₃** [8, 0]. A tangent drawn to the curve at u=0.8 cuts the X axis. Compute the point of intersection between the tangent and X axis

Solution:

Vector valued parametric equation of the curve is given by

$$P(u) = \sum_{i=0}^{3} B_{i,3}(u)P_i, \forall u \in [0,1]$$

The blending functions are

$$B_{0,3}(u) = (1-u)^3$$
$$B_{1,3}(u) = 3u(1-u)^2$$
$$B_{2,3}(u) = 3u^2(1-u)$$
$$B_{3,3}(u) = u^3$$

Substituting the values, the curve equation becomes

$$P(u) = [X(u) \ Y(u)] ; \ \forall u \in [0,1]$$

where

$$X(u) = 15u(1-u)^2 + 24u^2(1-u) + 8u^3$$
$$y(u) = 24u(1-u)^2 + 15u^2(1-u) \qquad \qquad \textbf{5.30}$$

Tangent vector to the curve is given by
$$P'(u) = [X'(u) \ Y'(u)] ; \ \forall u \in [0,1]$$

For **u=0.8**, the point on the curve (Equation 5.30) and the associated tangent vector are **P(u) = [7.648, 2.688]** , **P'(u) = [3.48,-11.52].**

The equation of the tangent is
$$P(v) = P(u) + v \, P'(u), \text{ where v is a scalar}$$
$$= [7.648 + 3.48 \, v, \ 2.688 - 11.52v]$$

At the point of intersection between the tangent and X-axis, y(v) = 0
Thus,

$$2.688 - 11.52v = 0$$
$$\textbf{v = 0.233}$$

The point of intersection between the tangent and the X axis is **[8.46, 0].**

Figure 5.26 show the Bezier curve and the tangent at **u=0.8**

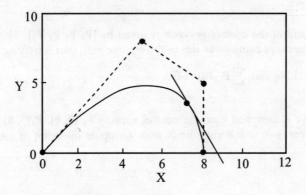

Figure 5.26 Bezier Curve

3. A cubic curve is to be designed which passes through the points **P₀[1, 1], P₁[6, 8], P₂[8, 5], and P₃[5, 5]** Derive the vector valued parametric equation of the curve.

Solution:

Vector valued parametric equation in the point vector form is given by

$$P(u) = \sum_{i=0}^{3} a_i u^i, \forall u \in [0,1]$$

$$= \begin{bmatrix} 1 & u & u^2 & u^3 \end{bmatrix} \begin{bmatrix} a_0 \\ a_1 \\ a_2 \\ a_3 \end{bmatrix}$$

Vector coefficients [a₀.........a₃] are to be determined for the boundary conditions outlined below

Let   u = 0; P(u) = P₀
            u = u₁; P(u) = P₁
            u = u₂; P(u) = P₂
            u = 1.0; P(u) = P₃

It can be seen that choosing different values of **u₁** and **u₂** in the parametric range, will produce a family of curves. As a typical case, choosing **u₁ = 1/3** and **u₂ = 2/3** Equation 5.29 becomes,

$$\begin{bmatrix} P_0 \\ P_1 \\ P_2 \\ P_3 \end{bmatrix} = \begin{bmatrix} 1 & 0 & 0 & 0 \\ 1 & u_1 & u_1^2 & u_1^3 \\ 1 & u_2 & u_2^2 & u_2^3 \\ 1 & 1 & 1 & 1 \end{bmatrix} \begin{bmatrix} a_0 \\ a_1 \\ a_2 \\ a_3 \end{bmatrix}$$

[P] = [M]. [A]

Thus,

$$[A] = [M]^{-1}[P] \qquad\qquad 5.31$$

Substituting terms in [M] and [P] in Equation 5.32 the coefficients matrix [A] can be computed

$$a_0 = [1, 1], \quad a_1 = [17.5, 49.0],$$
$$a_2 = [-4.5, -103.5], \quad a_3 = [-9.0, 58.5]$$

The parametric equation of the curve will be

$$P(u) = [X(u)\ Y(u)]\ ;\ \forall\ u\ \in [0,1]$$

where

$$X(u) = 1 + 17.5u - 4.5u^2 - 9.0u^3$$
$$y(u) = 1 + 49.0u - 103.5u^2 + 58.5u^3$$

Points on the curve can be computed by assigning various values to **u** $\in$ **[0 , 1]** .Figure 5.27 shows the curve and the points

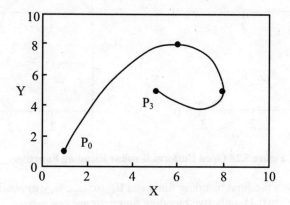

Figure 5.27 Cubic Curve through Points

4. An open uniform B-spline curve is designed from control points **[P₀ P₁ P₂ P₃]**.Show that for the order **k = 2**, the curve coincides with the control polygon.

Solution:

For open uniform B-spline curve with four control points (n+1= 4) and order k=2, the number of knots = 6. The open uniform knot vector is **T [0 0 1 2 3 3]**.

Using Cox-Deboor recursion relation, the blending functions **B$_{i,k}$(u)** can be computed.

For k =1,

$$B_{1,1}(u) = 0$$
$$B_{2,1}(u) = 1.0 \qquad\qquad 0 \le u < 1$$
$$B_{3,1}(u) = 1.0 \qquad\qquad 1 \le u < 2$$
$$B_{4,1}(u) = 1.0 \qquad\qquad 2 \le u < 3$$
$$B_{5,1}(u) = 0$$

For k =2,

$$B_{1,2}(u) = (1-u) \qquad 0 \le u < 1$$
$$= 0 \qquad \text{otherwise}$$
$$B_{2,2}(u) = u \qquad 0 \le u < 1$$
$$= (2-u) \qquad 1 \le u < 2$$
$$= 0 \qquad \text{otherwise}$$
$$B_{3,2}(u) = (u-1) \qquad 1 \le u < 2$$
$$= (3-u) \qquad 2 \le u < 3$$
$$= 0 \qquad \text{otherwise}$$
$$B_{4,2}(u) = (u-2) \qquad 2 \le u < 3$$
$$= 0 \qquad \text{otherwise}$$

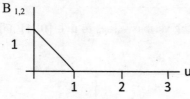
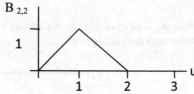
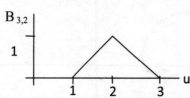
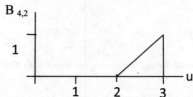

Figure 5.28 Open Uniform B-spline Blending Functions

Figure 5.28 shows the final blending functions $B_{1,2}(u)$....... $B_{4,2}(u)$ obtained. It can be seen that in any parametric range (say 0-1), only two blending functions are non zero.

To compute the shape of curve, say in range $u \in [0, 1]$, the curve equation is

$$P(u) = B_{1,2}.P_1 + B_{2,2}.P_2 + B_{3,2}.P_3 + B_{4,2}.P_4$$
$$= (1-u) P_1 + u P_2 \qquad \qquad \textbf{5.33}$$

This is, infact, a parametric equation of the straight line connecting P_1 and P_2. Same situation will hold for other parametric ranges viz [1, 2) and [2, 3)

Thus the B- Spline curve of order (k=2) coincides with the control polygon.

5.11 REVIEW QUESTIONS

1. Explain Why
   - For CAGD applications, parametric form of curve representation is preferred compared to the *Implicit/Explicit* types.
   - B-Spline curve offers more flexibility for shape control compared to Bezier curve
   - Hermite curve is not preferred for interactive curve synthesis.

2. Prove that a Bernstein Basis function $B_{i,n}(u)$ used in Bezier curve has maximum value at $u=i/n$ in the parametric range **[0,1].**

3. Two parametric curves $P(u) = [u, u^2, 0]$ and $Q(v) = [v, 0, -v^2]$, $u, v \in [-1, 1]$ are joined in space. Examine the conditions of continuity at the joint.

4. A Bezier curve designed from the control points $P_0[0, 0]$, $P_1[3, 5]$, and $P_2[5, 0]$ is clipped by a window having corner points $Q_1[1, 0]$, $Q_2[6, 0]$, $Q_3[6, 6]$ and $Q_4[1, 6]$. Compute the part of the curve contained in the window. Draw a neat sketch.

5. Using De Casteljau's subdivision algorithm, bisect the Bezier curve. Use control points in Example1 section 5.10. Verify the result.

6. An Open Uniform B-Spline curve is to be designed from six control points. The order of the curve is $K = 3$. Using Cox DeBoor relation, compute the blending functions. Draw a neat sketch showing the variation of the blending functions with parameter u.

Chapter 6

Design of Surfaces

Surfaces play an important role in the design of products. Today product designers create aesthetically smooth functional surfaces on several domestic and industrial products such as soaps, glassware, pottery, furniture, domestic appliances, automobiles, aircraft and ships. In computer aided geometric design (CAGD) scenario, designers need flexible tools to interactively create surfaces, tweak them into various shapes and blend them to create complex product shapes.

This chapter presents in details, the mathematical basis for interactive synthesis of surfaces used for CAGD. Suitable examples have been included.

6.1 TYPES OF SURFACES

A look at domestic and industrial products around us show that a wide variety of surfaces exists. These can be broadly classified into different categories such as planar surfaces, surfaces of revolution (wine glass, vases, shafts, gears), extruded surfaces (channels, I section beams, rails), complex sweep surfaces (screw threads) and freeform surfaces (aircraft wings, ship hulls, dies molds). In practice, a complex part surface like an automobile fender or dashboard, will comprise of multiple surface patches joined to each other to provide surface continuity. Design of such complex multi surfaces is important from the desired functional requirements of the part since the part (CAD) model will be used for application tasks like visualization (rendering), Computational Fluid Dynamic (CFD) analysis or part programming for the numerical control (CNC) machines.

6.2 REPRESENTATION OF SURFACES

A surface in 3D space can be conceived as the locus of a curve which moves as per a specified trajectory (path) for a limited distance. The resulting surface is a bounded geometric entity in $\mathbf{R}^3$ space. The nature of the curve and the trajectory path govern the type and complexity of the surface generated.

Traditionally a surface is represented in engineering drawing as a mesh of lines (curves) on the surface. In ship building and aircraft industry, the surfaces are represented in terms of cross sectional curves of the surface along a predefined axis. The resulting surface is an envelope of these cross sectional curves. This technique of surface design/ manufacture is termed as *Lofting*. When surfaces are known apriori such as the ones on existing parts or the styling clay models in automobile industry, the surface representation is in terms of 3D points obtained by inspection of parts using 3D coordinate Measuring Machines (CMM) or a Laser Scanner. In medical imaging applications such as CAT scan, MRI or X-ray radiography, the organ surfaces are represented as points along specific cross sectional contours.

Though the representation of surfaces in terms of points, cross sectional curves or net of lines is simple, it suffers from some major limitations. In absence of a surface equation, it is difficult to compute intermediate points/ lines (curves) on the surface. Further the surface properties like tangent, normal, curvature as well as surface intersections with other surfaces/ planes/ curves cannot be computed. These properties are particularly important in various CAD/CAM applications such as surface rendering, CFD analysis and CNC part programming to name a few. It is therefore, important to represent surfaces in mathematical form for CAGD.

Mathematical basis for surface synthesis and shape manipulation will be presented in the sections to follow.

6.3 MATHEMATICAL BASIS FOR SURFACE REPRESENTATION

Surfaces can be primarily represented in terms of *Implicit* functions or *Parametric* functions. Implicit surface equations are typically of the form **f (x, y, z) = 0** where the function **f( )** decides the degree and complexity of the surface. Parametric surfaces on the other hand, are defined in terms of parameters, **u, v** (scalars) and have the form of equation as **[X (u, v), Y (u, v), Z(u, v]** where X(), Y(), Z() are the parametric functions governing the degree/ shape of the surface. It is well known that all implicit functions cannot be represented into parametric form. However all parametric forms can be converted to implicit form by eliminating the parameters. Compared to the parametric form, implicit surfaces are more versatile in the representation of shapes. However they too have limitations.

6.3.1 Implicit Surfaces

From a mathematical point of view, implicit surfaces are two dimensional geometric shapes that exist in 3D space. The surface can be conceived to be an infinitesimally thin band of some quantity to be represented (say stress, temperature, pressure) which varies within the volume but is constant along the surface.

Mathematically the implicit surface comprises of points in 3D which satisfy some particular relationship say **f (p) = 0. P** is on the surface. If **f (p) < 0,** the point P is on one side (normally inside) of the surface. The relation **f ()** will decide the form of function and in turn, the shape of the surface in geometric design.

For CAGD applications, an implicit surface equation is typically of the form **f(x, y, z) = 0** i.e. a set of points $\{P \in R^3 ; f(p) = 0\}$. **f ()** is a typical mathematical function, (expression). For example if **f** () is a polynomial, the resulting surface is termed as an *Algebraic Surface*. In CAGD applications, implicit surfaces of degree 2, termed as *Quadrics* are frequently used.

A generalized equation of a second degree Quadric surface is given by

$$ax^2 + by^2 + cz^2 + 2f\,yz + 2g\,zx + 2h\,xy + 2px + 2qy + 2yz + d = 0 \qquad \textbf{6.1}$$

Examples of quadric surfaces include Cone, Cylinder, Ellipsoid, Elliptic cone, Elliptic cylinder, Elliptic Hyperboloid, Elliptic Paraboloid, Hyperbolic cylinder, Hyperbolic Paraboloid, and sphere. A quadric surface intersects a plane in a Conic section. Implicit equations for typical quadric surfaces are listed in Table 6.1.

TABLE 6.1
Implicit Surface Equations

Surface	Equation
Ellipsoid	$\dfrac{x^2}{a^2}+\dfrac{y^2}{b^2}+\dfrac{z^2}{a^2}=1$
Elliptic Cone	$\dfrac{x^2}{a^2}+\dfrac{y^2}{b^2}-\dfrac{z^2}{c^2}=0$
Elliptic Paraboloid	$z=\dfrac{x^2}{a^2}+\dfrac{y^2}{b^2}$
Hyperbolic Paraboloid	$z=\dfrac{y^2}{b^2}-\dfrac{x^2}{a^2}$
Hyperboloid (one sheet)	$\dfrac{x^2}{a^2}+\dfrac{y^2}{b^2}-\dfrac{z^2}{c^2}=1$

Figure 6.1 shows some typical quadric (Implicit) surfaces.

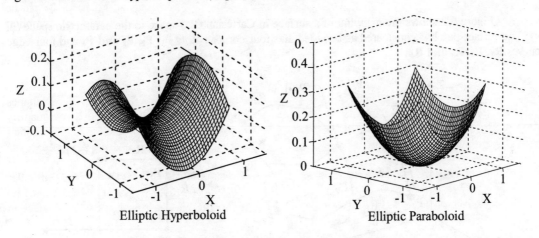

Elliptic Hyperboloid   Elliptic Paraboloid

Figure 6.1 Implicit (Quadric) Surfaces

Implicit surface representations have some inherent advantages and limitations.

Advantages

- Implicit surfaces are versatile in representing shapes of varying complexities.
- It is quite easy to compute and test if a point lies on the surface, inside or outside.
- Computing surface normal, tangent, surface intersections and Boolean operations is fairly simple.
- It is easy to handle topological changes to model deformable objects.

Limitations

- Implicit surface equations give indirect specification of the surface compared to the parametric form which is more directly perceived by the designer.
- It is very difficult to describe and model sharp features. Implicit surfaces are more suitable for rounded (blob) objects.
- Rendering of surfaces is a difficult task needing special algorithms. Speed of rendering is thus, slow.

In CAGD, parametric form of representation of curves and surfaces is widely used due to several advantages offered by them. These are discussed below.

6.3.2 Parametric Surfaces.

Parametric surface equations essentially map a point **P (x, y, z)** in $\mathbf{R}^3$ space to the parametric space **(u, v)**. Important advantages of using parametric form of representation are axis independence, affine invariance and ease of computing points on the surface. In this book, parametric form of surface representation will be followed throughout.

Figure 6.2 shows the mapping of a surface in Cartesian $(\mathbf{R}^3)$ space to the parametric space (u, v). The surface patch is considered as bounded, has four corner points $\mathbf{P_1, P_2, P_3}$ **and** $\mathbf{P_4}$ and four edge curves $\mathbf{R_1, R_2, R_3}$ **and** $\mathbf{R_4}$.

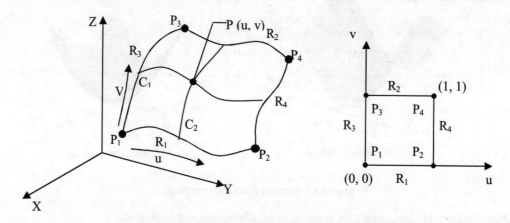

Figure 6.2 Parametric Surface Representation

Conceptually the surface can be considered to have *Infinite* number of lines/ curves (mesh). Without loss of generality, they could be considered as lines along two parameters **u** and **v** where **u, v** are scalars. Normalizing the scalars **u, v** ∈ **[0,1]** would lead to a rectangular (square) parametric space (Figure 6.2). The surface patch is mapped into the parametric space such that the four corner points coincide with the vertices of the parametric space. In particular, the points will be $\mathbf{P_1}$ **(0, 0)**, $\mathbf{P_2}$ **(1, 0)**, $\mathbf{P_3}$ **(1, 1)** and $\mathbf{P_4}$ **(0, 1)**. The edge curves will be *Isoparametric* curves for which one of the parameters (u or v) will be constant and the other varies between 0 to 1. For example, the edge curve R_1 will have v = 0 and u ∈ **[0, 1]**.

Any point on the surface P(x, y, z) can be conceived to be the intersection of two isoparametric curves $C_1(u)$, $C_2(v)$ for which the other parameter (v or u) will be constant. (Figure 6.2) The vector valued parametric equation of the surface is

$$P(x,y,z) = P\big[x(u,v),\ y(u,v),\ z(u,v)\big], \forall u,v \in [0,1] \qquad \textbf{6.2}$$

X (u, v), Y (u, v), Z (u, v) are mapping functions which decide the degree and shape of the resulting surface. Since two independent (scalar) parameters **u** and **v** are used, the surface is termed as a Bi- Parametric surface patch.

Mathematical basis for the design of various types of surfaces will be presented in the sections to follow.

6.4 DESIGN OF SWEEP SURFACES

Surfaces created by sweeping a curve in R^3 space along a chosen trajectory are termed as *Sweep* surfaces. The trajectory path can be any standard geometric transformation or some known curve. For example typical curve in X-Y plane when rotated about X axis can generate a surface of revolution while translation along Z can generate a extruded (tabulated) surface. These are, in essence, *Rotational* and *Translational* sweep operations.

6.4.1 Rotational Sweep Surfaces.

Rotational sweep surfaces are widely used on a variety of products. These typically include vases, wine glasses, pottery made on potters wheel, parts produced on lathe machines to name a few.

Without loss of generality, it is considered here that the basic curve to be swept lies in the X-Y plane and is rotated about X axis in a counterclockwise direction. Figure 6.3 shows the curve and the sweep (rotation) direction. The basic curve is represented as a vector valued parametric equation.

$$P(u) = \big[x(u)\ \ y(u)\ \ 0\big], \forall u \in [0,1] \qquad \textbf{6.3}$$

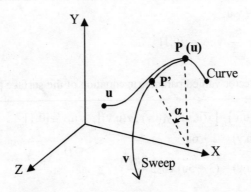

Figure 6.3 Rotational Sweep operation

Parameter **v** is considered along the sweep direction (angle α). During rotation by an angle α around X axis, a typical point on the curve **P[x, y] = [x(u) y(u) 0]** gets transferred to **P'** which lies on the swept surface.

The vector valued parametric equation of the surface will be

$$P(u,v) = [x(u,v), y(u,v), z(u,v)], \forall u,v \in [0,1]$$

where,
$$x(u,v) = x(u)$$
$$y(u,v) = y(u). \cos\alpha$$
$$z(u,v) = y(u). \sin\alpha \qquad\qquad 6.4$$

Fixing extents of u and v in R^3, a bounded surface will be designed. Few illustrative examples are included here.

Sweeping a Line

Let the basic curve in X-Y plane be a straight line through points $P_1(x_1\ y_1)$ and $P_2(x_2\ y_2)$. It is rotated about X axis in counterclockwise direction by angle α. Depending upon the orientation of the line P_1P_2 with X axis, various shapes of surfaces will occur. For example a Planar Disc will result for P_1P_2 parallel to Y axis, a Cylindrical Surface for P_1P_2 parallel to X axis and a Cone Surface for P_1P_2 intersecting with X axis.

As an illustration, let **P₁(2,3)** and **P₂(8,8)**. The equation of the basic curve (line) is

$$P(u) = P_1 + u(P_2 - P_1), \quad \forall u \in [0,1]$$
$$P(u) = [x(u)\ y(u)\ 0]$$

where,
$$x(u) = 2 + 6u$$
$$y(u) = 3 + 5u \qquad\qquad 6.5$$

Let the base curve be rotated by 120 degrees to create the surface. Parameterizing the sweep direction,
$$\alpha = v.\alpha_{max}$$
$$= \frac{2\pi}{3}.v \ ; v \in [0,1]$$

Using Equation 6.5, the vector valued parametric equation of the surface patch is

$$P(u,v) = [x(u,v), y(u,v), z(u,v)], \quad \forall u,v \in [0,1]$$

where,
$$x(u,v) = 2 + 6u$$

$$y(u,v) = (3 + 5u).\cos\left(\frac{2\pi v}{3}\right)$$

$$z(u,v) = (3 + 5u). \sin\left(\frac{2\pi v}{3}\right) \qquad\qquad 6.6$$

Figure 6.4 shows the conical surface represented by Equation 6.6

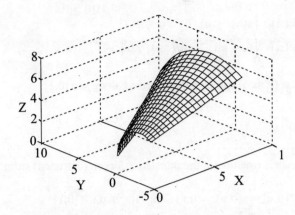

Figure: 6.4 Rotational Sweep – Cone Surface

The basic curve can be chosen to be any curve like a conic curve, Bezier, B-Spline or NURB curve to obtain the desired swept surface. The surface equation (Equation 6.6) is axis independent and hence any affine transformation (Chapter 3) can be applied to the resulting surface.

6.4.2 Translational Sweep Surfaces

Translational sweep essentially extrudes the basic curve along a chosen path (direction). The trajectory path can be any direction in R^3 space. Care should be taken to see that the swept curve does not degenerate to some point or create an absurd shape.

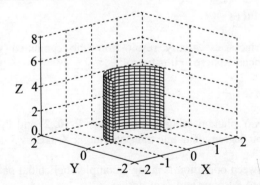

Figure 6.5 Translational Sweep – Cylindrical surface

Figure 6.5 shows a basic curve in X-Y plane extruded along Z axis. As an example, let the basic curve be a semicircular arc of an origin centered unit circle extruded along Z-axis by 5 units. Parameterizing θ,

$$\theta = u\,\theta_{max}$$
$$= \pi\,u$$

The equation of the base curve is

$$P(u) = [x(u) \quad y(u) \quad 0], \forall u \in [0,1]$$
$$P(u) = [\cos(\pi u), \quad \sin(\pi u), \quad 0]$$

The direction of extrusion (Z axis) can be parameterized along $v \in [0,1]$.

> Thus
$$Z = v \, Z_{max}$$
$$= 5v$$

The vector valued parametric equation of the translation sweep (extruded) surface is

$$P(u,v) = [x(u,v), y(u,v), z(u,v)], \quad \forall u,v \in [0,1]$$

Where, $x(u,v) = x(u) = \cos(\pi u)$

$y(u,v) = y(u) = \sin(\pi u)$

$z(u,v) = 5v$ 6.7

Figure 6.5 shows the cylindrical surface represented in Equation 6.7 Again as before, any curve (conic, Bezier, B-Spline, NURBS) can be used as the basic curve for sweep operation.

Translation sweep creates surfaces which have the same cross section (basic curve) at various sections along its length. Such surfaces are widely seen on products such as extruded angles/ channels, rails used on railway tracks, corrugated sheets, columns and beams used in civil construction.

6.4.3 Hybrid Sweep Surfaces

Hybrid sweep surfaces essentially result from the combined geometric transformation operations. An example is included here to illustrate

Complex Sweep Operation

A straight line in X-Y plane passing through points **P₁ (0, 2)** and **P₂ (1, 0)** is rotated along X-axis counterclockwise by 360 degrees. The line translates along X-axis by 3 units simultaneously.

It is seen that the sweep operation is along a complex helicoidal path along X-axis. The path can be conceived as combined *Rotation* and *Translation*.

The basic curve is represented as

$$P(u) = P_1 + u(P_2 - P_1), \quad \forall u \in [0,1]$$
$$= [x(u) \quad y(u) \quad 0]$$

Where, $X(u) = u$

$Y(u) = 2(1-u)$

Sweep causes rotation by angle $\alpha = 2\pi$ and the translation along X by 3 units,
Parameterizing along v,

$$\alpha = v.\alpha_{max}$$
$$= 2\pi v; \, v \in [0,1]$$

The incremental motion along X axis during the rotation by α will be

$$\Delta x = \frac{\alpha}{\alpha_{max}}.(\text{pitch of helix})$$
$$= 3v$$

The vector valued parametric equation of the surface generated will be

$$P(u,v) = [x(u,v), \, y(u,v), \, z(u,v)], \forall \, u,v \in [0,1]$$

where

$$x(u,v) = x(u) + \Delta x$$
$$= u + 3v$$
$$y(u,v) = 2(1-u)\cos(2\pi v)$$
$$z(u,v) = 2(1-u)\sin(2\pi v) \qquad \qquad \textbf{6.8}$$

Figure 6.6 shows the shape of the complex surface generated. It is a helicoidal surface commonly occurring on the flanges of screw threads. As before, any basic curve can be used to generate more complex surface shapes.

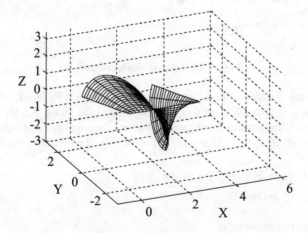

Figure 6.6 Helicoidal Sweep

6.5 DESIGN OF SURFACE PATCHES

Many a times the intended surface is too complex to be modeled as a single surface. Typical examples could be the surfaces in automobiles, aircrafts or ships. In such cases, the designer follows an approach similar to making quilts using patchwork. The complex surface is designed in terms of smaller surface patches which are *stitched* together to give position and tangent continuities. A typical car body surface designed this way may have thousands of surface patches. The patches provide flexibility to the designer to locally tweak the surface to suit aesthetics and functionality.

Figure 6.2 shows the mapping of a typical surface patch into the bi-parametric space (u, v). The surface patch in R^3 space passes through four corner points and has four defined edge curves. It is a bounded surface. The shape and complexity (degree) of the patch depends upon the boundary conditions during the modeling.

Mathematical basis for the modeling and shape control of various surface patches will be presented

6.5.1 Bilinear surface Patch

A bilinear surface patch is characterized by the fact that all the four edge curves as well as curves on the patch are straight lines. Figure 6.7 shows the bilinear surface patch in R^3 and its mapping in parametric space.

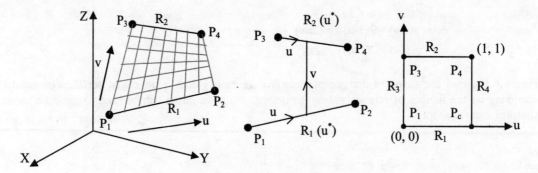

Figure 6.7 Bilinear Surface Patch

The patch passes through four corner points P_1, P_2, P_3 and P_4 and has four linear edges R_1, R_2, R_3 and R_4 as shown. It is considered that the parameter **u** is along R_1, R_2 while parameter **v** is along R_3, R_4. To construct the surface patch, without loss of generality, it is considered that the edge curves R_1 and R_2 are constructed first (Figure 6.7). Vector valued parametric equation of curve R_1 is

$$R_1(u) = P_1 + u(P_2 - P_1), \quad \forall u \in [0,1]$$
$$= (1-u)\,P_1 + u\,P_2 \tag{6.9}$$

In a similar way, equation of curve R_2 is

$$R_2(u) = P_3 + u(P_4 - P_3), \quad \forall u \in [0,1]$$
$$= (1-u)\,P_3 + u\,P_4 \tag{6.10}$$

The patch can be constructed by defining *iso-parametric* line along **v** for corresponding chosen values of $u = u^*$. Thus the equation of the isoparametric line along v will be

$$P(u^*,v) = R_1(u^*) + v\,(R_2(u^*) - R_1(u^*))$$
$$= (1-v)\,R_1(u^*) + v R_2(u^*) , \quad \forall\, v \in [0,1] \tag{6.11}$$

$\mathbf{u}^*$ has been arbitrarily chosen for $u \in [0,1]$ and can be dropped for future. From Equation 6.8 and Equation 6.10, the vector valued parametric equation of the bilinear surface patch is

$$P(u,v) = (1\text{-}u)(1\text{-}v)P_1 + u(1\text{-}v)P_2 + (1\text{-}u)v\,P_3 + uv\,P_4, \quad \forall \quad u,v \in [0, 1] \qquad \textbf{6.12}$$

The above equation can be written in the matrix form as

$$P(u, v) = \begin{bmatrix} 1-v & v \end{bmatrix} \begin{bmatrix} P_1 & P_2 \\ P_3 & P_4 \end{bmatrix} \begin{bmatrix} 1-u & u \end{bmatrix}^T \qquad \textbf{6.13}$$

From Equation 6.12 and 6.13, it is seen that a point on the bilinear surface is the weighted average of the corner points. Writing Equation 6.13 in this form,

$$P(u,v) = \sum_{i=1}^{4} B_i(u,v).P_i$$

where

$$B_1(u, v) = (1\text{-}u)(1\text{-}v)$$
$$B_2(u, v) = u(1\text{-}v)$$
$$B_3(u, v) = (1\text{-}u)v$$
$$B_4(u, v) = uv \qquad \textbf{6.14}$$

From Equation 6.14 it can be seen that each blending function has the maximum value of 1.0 at the respective corner point of the parametric space and zero at the other corner points. This property ensures that the surface patch passes through the corner points P_1, P_2, P_3 and P_4. By choosing proper corner points, surface patches of various shapes can be synthesized Figure 6.8 shows some typical bilinear surface patches obtained by changing the control points.

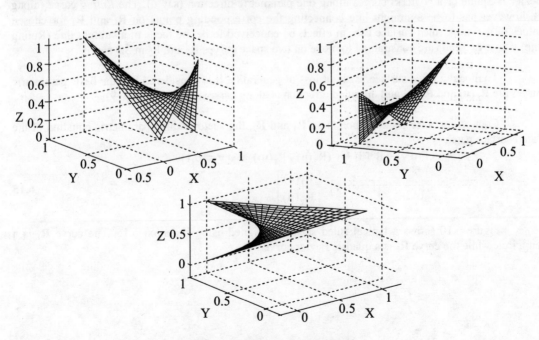

Figure 6.8 Bilinear Surface Patches

It can be seen from Figure 6.8 that the bilinear surface patch has the capability to model *Twisted* surfaces, though each parametric curve on the surface will be a straight line.

6.5.2 Ruled Surfaces

Ruled surface is a more general case of the bilinear surface patch. Figure 6.9 shows the concept of a ruled surface and its mapping in the parametric space.

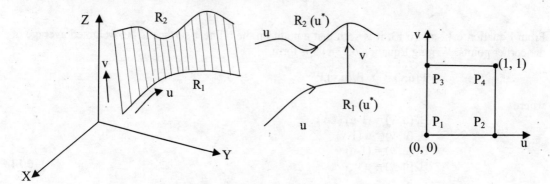

Figure 6.9 Ruled Surface Patch

Compared to a bilinear surface patch, a Ruled surface patch has opposite edge curves which need not be straight lines. For example, the curve R_1 and R_2 can be straight lines, conic curves, a Bezier B-Spline or a NURBS curves along one parametric direction (say u). The *Ruling* curve (along v), however has to be a straight line connecting the corresponding points on R_1 and R_2 for chosen values of $\mathbf{u} = \mathbf{u}^*$. A ruled surface can, in effect, be conceived to be the locus of a straight line (Ruling line) whose two ends are constrained to move on two space curves R_1 and R_2 in 3D space.

To model a ruled surface, without loss of generality, it is considered that the basic parametric curves are $R_1(u)$ and $R_2(u)$ and the ruled direction is along parameter $\mathbf{v}$. (Figure 6.9)

Connecting corresponding points on R_1 and R_2, the vector valued parametric equation of the *Ruled* surface patch is

$$P(u,v) = R_1(u) + v\,(R_2(u) - R_1(u)),\ \forall\ u,v \in [0,1]$$

$$= \begin{bmatrix} 1-v & v \end{bmatrix} \begin{bmatrix} R_1(u) \\ R_2(u) \end{bmatrix} \qquad \textbf{6.15}$$

Figure 6.10 shows a typical ruled surface designed using Equation 6.15. The curve R_1 is a semicircle while the curve R_2 is a quadric Bezier curve.

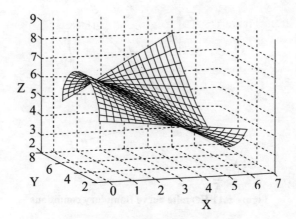

Figure 6.10 Ruled Surface

Ruled surfaces are widely used in designing parts of aircrafts (wings), ships (hulls), container bottle and in medical imaging applications viz. organ modeling from CAT scan, MRI, X-ray data.

6.5.3 Coon's Surface Patch

In his pioneering work, Coon developed a methodology to design complex surface patch from known geometric boundary conditions. Gorden independently developed a similar procedure terming it as *transfinite interpolation*. Forrest proposed a simplified version of Coons patch with restricted boundary conditions.

In essence, the Coon's methodology enables the design of a complex surface patch passing through four corner points which are connected by edge curves designed to a required degree. Coons patch is also termed as a *Four Curve* patch as it interpolates a surface between the four known boundary curves.

Mathematical basis for the design of Coon's surface patch will be discussed.

Design of Coon's Surface Patch

Parametric surface patch design philosophy shown in Figure 6.2 essentially applies to a Coon's surface patch also. The patch passes through four corner vertices P_1, P_2, P_3 and P_4 and has four edge curves R_1, R_2, R_3 and R_4. Without loss of generality, the edges R_1 and R_2 are considered to be isoparametric curves with parameter **u** while R_3 and R_4 along parameter **v**. Mathematical basis for the design of edge curves for Coon's patch is Hermite curves (Chapter 5). To explain the concept, it is considered here that all the four curves R_1, R_2, R_3 and R_4 are parametric cubic Hermite curves. The Coon's surface patch will thus, be a Bi cubic one.

Boundary Conditions

A cubic Hermite curve needs four boundary conditions for design. These include position vectors and tangent vectors at the two ends of the curve (Figure 6.11)

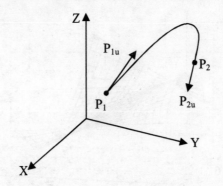

Figure 6.11 Hermite curve Boundary conditions

Specifically

at   u=0 $P(u) = P_1$, $P'(u) = P_{1u}$
       u=1 $P(u) = P_2$, $P'(u) = P_{2u}$

where tangent vectors are partial derivatives along **u**. Thus,

$$P_{1u} = \frac{\partial P_1}{\partial u} \ , P_{2u} = \frac{\partial P_2}{\partial u}$$

A similar situation will exist for curve design along v. Figure 6.12 shows a typical Coons bicubic surface patch. It is seen that each point **P(u, v)** on the surface is the intersection of two isoparametric space curves say $C_1(u)$ and $C_2(v)$. The same applies to the four corner vertices. In particular, corner point P_1 is the intersection of $R_1(u)$ and $R_3(v)$ at u=v=0.

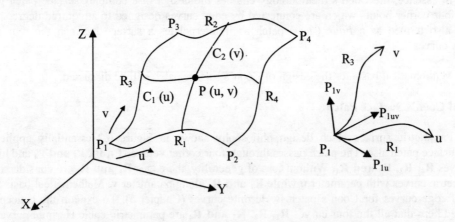

Figure 6.12 Coons Surface Patch Boundary Conditions

Coon's surface is the tensor product of two cubic isoparametric curves (along u and v) each needing four boundary conditions. Coon's Surface patch will thus, need 16 (4 x 4) boundary conditions. At each corner vertex, two tangent vectors exist as shown in Figure 6.12. For example for P_1, the tangent vectors will be the partial derivatives,

$$P_{1u} = \frac{\partial P(u,v)}{\partial u} \Bigg|_{u=0, \, v=0}$$

$$P_{1v} = \frac{\partial P(u,v)}{\partial v} \Bigg|_{u=0, \, v=0}$$

There will be 8 tangents vectors for the four corner points. Four additional boundary conditions were proposed by Coons in the form of *Twist vectors* viz.

$$P_{uv} = \frac{\partial^2 P(u,v)}{\partial u \, \partial v}$$

The 16 boundary conditions for the design of Coons Bi cubic surface patch are as follows.

    4 Corner Points   : P_1, P_2, P_3, P_4

    8 Tangent Vectors   : $P_{1u}, P_{2u}, P_{3u}, P_{4u}$

                                      $P_{1v}, P_{2v}, P_{3v}, P_{4v}$

    4 Twist vectors   : $P_{1uv}, P_{2uv}, P_{3uv}, P_{4uv}$

Mathematical Equation

Figure 6.13 shows the strategy for the modeling of Coons surface patch. Without loss of generality, it is considered that the edge curves $R_1(u)$ and $R_2(u)$ are constructed first and then corresponding points on them for say $u = u^*$ are joined by a parametric curve along v.

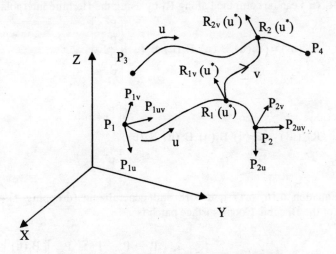

Figure 6.13 Modeling of Coons Surface Patch

Vector valued parametric equation of the cubic Hermite curve $R_1(u)$ is given by

$$R_1(u) = \begin{bmatrix} B_1(u) & B_2(u) & B_3(u) & B_4(u) \end{bmatrix} \begin{bmatrix} P_1 \\ P_2 \\ P_{1u} \\ P_{2u} \end{bmatrix} \qquad \text{6.16}$$

The Hermite cubic blending functions (Chapter 5) are given by

$$B_1(u) = 1 - 3u^2 + 2u^3$$
$$B_2(u) = 3u^2 - 2u^3$$
$$B_3(u) = u - 2u^2 + u^3$$
$$B_4(u) = -u^2 + u^3$$

In a similar manner, equation for **R₂(u)** is

$$R_2(u) = \begin{bmatrix} B_1(u) & B_2(u) & B_3(u) & B_4(u) \end{bmatrix} \begin{bmatrix} P_3 \\ P_4 \\ P_{3u} \\ P_{4u} \end{bmatrix} \qquad \text{6.17}$$

To construct a cubic Hermite curve along v from $R_1(u^*)$ and $R_2(u^*)$ (Figure 6.20) corresponding to $u = u^*$, the following boundary conditions will be needed.

$$
\begin{aligned}
&\text{2 corner Points} &&: R_1(u^*), R_2(u^*) \\
&\text{2Tangent Vectors} &&: R_{1v}(u^*), R_{2v}(u^*) \\
&\text{(along v)}
\end{aligned}
$$

The tangent vector $R_{1v}(u^*)$ can be computed along R_1 by using the Hermite interpolation. In particular,

$$R_{1v}(u^*) = \begin{bmatrix} B_1(u) & B_2(u) & B_3(u) & B_4(u) \end{bmatrix} \begin{bmatrix} P_{1v} \\ P_{2v} \\ P_{1uv} \\ P_{2uv} \end{bmatrix}$$

Similarly

$$R_{2v}(u^*) = \begin{bmatrix} B_1(u) & B_2(u) & B_3(u) & B_4(u) \end{bmatrix} \begin{bmatrix} P_{3v} \\ P_{4v} \\ P_{3uv} \\ P_{4uv} \end{bmatrix} \qquad \text{6.18}$$

Substituting from Equation 6.16, 6.17 and 6.18 and generalizing (dropping *), the vector valued parametric equation of the Bi cubic Coons surface patch is

$$P(u, v) = \begin{bmatrix} B_1(v) & B_2(v) & B_3(v) & B_4(v) \end{bmatrix} \begin{bmatrix} P_1 & P_2 & P_{1u} & P_{2u} \\ P_3 & P_4 & P_{3u} & P_{4u} \\ P_{1v} & P_{2v} & P_{1uv} & P_{2uv} \\ P_{3v} & P_{4v} & P_{3uv} & P_{4uv} \end{bmatrix} \begin{bmatrix} B_1(u) \\ B_2(u) \\ B_3(u) \\ B_4(u) \end{bmatrix} \qquad \text{6.19}$$

Knowing the boundary conditions **[P]**, any point on the surface patch can be computed for $\forall\; u,v \in [0,1]$. The boundary condition matrix **[P]** is quite neat and symmetric as under

$$\begin{bmatrix} P & P_u \\ P_v & P_{uv} \end{bmatrix}_{4\times 4}$$

Each quadrant is a 2 x 2 subset comprising either Position vectors, its partial derivatives (u,v) or the twist vectors. Forrest simplified Coons bi cubic patch by considering all Twist vectors to be zero ($P_{uv} = 0$). This condition creates surface patch having *Flat* ends at the corner points.

6.5.4 Bezier, B- Spline and NURBS surface patches

Though Coon's surface patch provides a mathematically robust and elegant solution to surface design, it has limitations similar to the Hermite curves. To design a Coon's surface patch, user needs to input boundary conditions comprising of 8 tangent vectors and 4 twist vectors. These are very abstract for a designer to think apriori. This difficulty is removed in the strategy for design of Bezier, B-Spline or NURBS surface patches, where in the user needs to input a mesh of control points only. The surface generally follows the shape of the mesh created by the control points. The user can thus, flexibly tweak the surface to suit his intent by adjusting the control points.

Mathematical basis for the design of Bezier, B-Spline or NURBS surface patches is identical except for the blending (basis) function which is Berstein, B-Spline or Rational as the case may be. As a result, the mathematical basis for the design of Bezier surface is explained here in detail.

Design of Bezier Surface patch

From a mathematical point of view, a Bezier surface is obtained by the tensor product of two Bezier space curves defined along parametric directions **u** and **v**. Figure 6.14 shows a typical Bezier surface patch in R^3 space and its mapping in the parametric (uv) space.

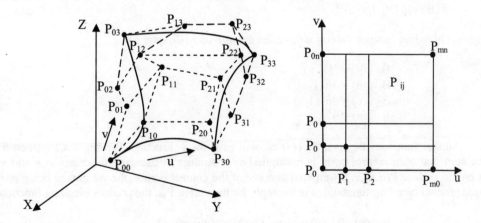

Figure 6.14 Modeling Bezier Surface Patch

Vector valued parametric equation of the surface patch is

$$P(u,v) = \sum_{i=0}^{m} \sum_{j=0}^{n} B_{i,m}(u) . B_{j,n}(v) . P_{ij}, \quad \forall u,v \in [0,1]$$

where $B_{i,m}(u)$, $B_{j,n}(v)$ are the Bernstein basis functions given by

$$B_{i,m}(u) = {}^{m}C_i \, u^i (1-u)^{m-i}$$
$$B_{j,n}(v) = {}^{n}C_j \, v^j (1-v)^{n-j}$$

6.20

To design the surface patch, a mesh of $(m+1)$ x $(n+1)$ control points (P_{ij}) will be required to be inputted by the designer. Figure 6.14 show the control mesh required for a typical patch. Without loss of generality, it is considered that there are $(m+1)$ control points along **u** direction giving the degree of Bezier curve as **m**. In a similar manner, the degree of the curve along **v** direction is **n** needing $(n+1)$ control points. It can be seen that the patch passes through the four corner points [P_{00}, P_{03}, P_{30}, P_{33}]. The control points on the edges of the mesh such as points P_{10}, P_{20} etc govern the edge curves while the control points in the centre of the mesh [P_{11}, P_{21}, P_{12}, P_{22}] govern the central shape (Hump) of the surface patch.

To design a Bicubic Bezier surface, let u= v=3. Vector valued parametric equation of Bicubic Bezier patch is

$$P(u,v) = \sum_{i=0}^{3} \sum_{j=0}^{3} B_{i,3}(u) . B_{j,3}(v) . P_{ij}, \quad \forall u,v \in [0,1]$$

Writing in the matrix form, the equation is

$$P(u,v) = [B_{0,3}(v) \quad B_{1,3}(v) \quad B_{2,3}(v) \quad B_{3,3}(v)] \begin{bmatrix} P_{00} & P_{10} & P_{20} & P_{30} \\ P_{01} & P_{11} & P_{21} & P_{31} \\ P_{02} & P_{12} & P_{22} & P_{32} \\ P_{03} & P_{13} & P_{23} & P_{33} \end{bmatrix} \begin{bmatrix} B_{0,3}(u) \\ B_{1,3}(u) \\ B_{2,3}(u) \\ B_{3,3}(u) \end{bmatrix}$$

$$= [B(v)] . [P] . [B(u)]^T$$

6.21

The typical blending functions along **u** direction are as under (Chapter 5)

$$B_{0,3}(u) = (1-u)^3$$
$$B_{1,3}(u) = 3u(1-u)^2$$
$$B_{2,3}(u) = 3u^2(1-u)$$
$$B_{3,3}(u) = u^3$$

Similar functions $B_{0,3}(v)$, $B_{1,3}(v)$ etc will exist for **v** direction. Writing out Equation 6.21, it can be seen that each control point is multiplied by a product of blending functions in **u** and **v**. The point on the surface **P(u,v)** is a weighted average of the control points; the weightage being provided by these product blending functions. For example for the vertex P_{00}, the product blending function is

$$B_{0,3}(u) . B_{0,3}(v) = (1-u)^3 (1-v)^3 \quad \forall u,v \in [0,1]$$

It is seen that the product blending function $B_{0,3}(u) . B_{0,3}(v)$ is **1** at **u=v=0** and vanishes at other corner points. A similar situation will exist at other corner points P_{30}, P_{03}, and P_{33}. The surface patch will thus, pass through the corner point P_{00}.

Figure 6.15 shows a typical bicubic Bezier surface patch.

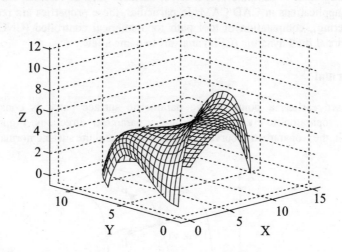

Figure 6.15 Bicubic Bezier Surface

User can interactively change the locations of the control points to change the shape of the surface which passes through the 4 corner points.

Important characteristics of Bezier surface are summarized below

- The surface is definable in terms of a mesh of control point. The surface generally follows the mesh, thus capturing *Designers intent*.
- Degree of the surface can be varied along the parametric directions (u,v), thus permitting versatility of shape synthesis.
- Degree of the curve in each parametric direction is one less than the number of control polygon vertices in that direction.
- The surface provides continuity of one less than the degree in that parametric direction (c^{m-1}, c^{n-1})
- The surface lies within the convex hull of the defining polygon. This is primarily due to the convex Hull property of the blending functions.
- The surface passes through the corner points. The other points of the control point mesh provide surface manipulation capability.
- The surface is invariant under an affine transformation.

The Bezier, B-spline or NURBS surfaces with their appropriate basis functions provide a very flexible and interactive tool to the designer for the synthesis and manipulation of surfaces in an ab initio design environment. Use of B-Spline and Rational basis functions provide further advantages of *Local control* and versatility in shape representation as was discussed in Chapter 5.

6.6 SURFACE PROPERTIES

Parametric equations discussed in earlier sections enable a designer to interactively create a surface patch, tweak its shape and compute points on the surface to display it as a net (mesh) of lines/ curves. It is often important to compute properties of surfaces such as Normal, Curvature to suit the intended downline applications in CAD/CAM. In particular, these properties are required for surface visualization (rendering), computation of tool path for numerical controlled (CNC) machines, flow studies in computational fluid dynamic (CFD) analysis to name a few.

6.6.1 Surface Normal

As discussed earlier, a point on the parametric surface patch is considered to be the intersection of two isoparametric curves along u and v parameter directions. Figure 6.16 shows the typical surface patch, the isoparametric curves, tangent vectors and the surface normal.

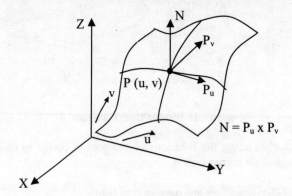

Figure 6.16 Tangent Vectors and Surface

The tangent vectors along these isoparametric curves are in essence, the partial derivatives computed at the point. Thus $P_u = \dfrac{\partial\ P(u,v)}{\partial u}$ and $P_v = \dfrac{\partial\ P(u,v)}{\partial v}$

Normal vector to the surface patch at the point P(u, v) is the cross product of the tangent vector P_u and P_v. Thus

$$N = \frac{P_u \times P_v}{|P_u \times P_v|}$$

6.22

6.6.2 Curvature and Shape of Surfaces

Curvature is an important property which tells about the local shape of the surface at the point.

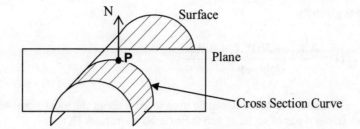

Figure 6.17 Surface Cross Section and Curvature

Figure 6.17 shows a typical surface showing a point P and the normal vector N at that point. A plane passing through P and containing the normal vector N will cut the surface to produce the curve of intersection. Depending upon the shape of the surface, the intersection curve could be convex, concave or flat (line) having positive, negative or zero curvature. Euler found that if the cutting plane is rotated around the normal N, there exist two mutually orthogonal directions along which the curvature is maximum and minimum (K_{min}, K_{max}). The directions of maximum and minimum curvatures can be computed from the equation of the surface. From computational geometry point of view, measures of *Gauss* and *Mean* curvatures are more important to infer about the shapes of surfaces. Gaussian curvature (K) is given by $\mathbf{K = K_{min} . K_{max}}$ while Mean curvature (H) is given by $\mathbf{H = \frac{|K_{min}| + |K_{max}|}{2}}$.

Dill and Rogers have given an elegant procedure for the computation of Gauss (k) and Mean (H) curvatures at a point from the parametric equation of a surface. It is explained below.

Vector valued parametric equation of the surface is given by
$$P(u,v) = [x(u,v), y(u,v), z(u,v)], \forall\, u,v \in [0,1]$$
Partial derivatives at a chosen point P(u, v) are computed. Using the notation,

$$P_u = \frac{\partial P(u,v)}{\partial u}, \quad P_v = \frac{\partial P(u,v)}{\partial v}$$

$$P_{uu} = \frac{\partial^2 P(u,v)}{\partial u^2}, \quad P_{vv} = \frac{\partial^2 P(u,v)}{\partial v^2}$$

$$P_{uv} = \frac{\partial^2 P(u,v)}{\partial u\, \partial v}$$

$$[A\ B\ C] = [P_u \times P_v] . [P_{uu}\, P_{uv}\, P_{vv}]$$

where **x** and **.** denote cross and dot vector products respectively.
Gaussian curvature is given by

$$K = \frac{AC - B^2}{|P_u \times P_v|^4} \qquad\qquad \textbf{6.23}$$

Mean curvature is given by

$$H = \frac{A|P_v|^2 - 2BP_u.P_v + B|P_u|^2}{2|P_u \times P_v|^3}$$

6.24

The sign of Gauss Curvature K can be used to infer about the local shape of the surface. Table 6.2 summarizes various types of surfaces and their Gauss curvatures (K)

TABLE 6.2
Types of surfaces and Gauss Curvature (K)

Gauss Curvature K	K_{min}	K_{max}	Types of surfaces
K > 0	+ -	+ -	Elliptic (Bump; Hollow)
K < 0	+ -	- +	Hyperbolic (Saddle)
K = 0	0 +/- 0	+/- 0 0	Cylindrical/Conical (Ridge, Hollow, Plane)

6.6.3 Developability of Surfaces

During the manufacture of products made from sheet /plates, it is often required to *Open out* or *Develop* the product shape on flat sheet without causing any tear/crack in the surface. Examples of such parts are cylindrical/conical pipes, pipe transition joints, ducts for air conditioning, parts of furnitures and domestic appliances, metal containers etc.

Gaussian curvature (K) is an important property which governs the developability of a surface. In order that a surface is *developable,* it must be *ruled.* If the surface has straight lines (rulings), the surface can be conceptually *opened* out by giving small rotations about these lines to lay on a flat sheet *without* tearing.

In order that a surface is developable, Gaussian Curvature (K) must be zero at all points on the surface. Thus Cone and Cylinder are developable surfaces while a sphere is not. All developable surfaces are *ruled* but the converse is not true.

6.7 EXAMPLES

1. An ellipse in X-Y plane has centre (10, 10) and semi major and semi minor axes 3 and 2 respectively. It is rotated about X axis in counterclockwise direction by 270 degrees. Derive vector valued parametric equation of the swept surface generated. Compute point on the surface at u= 0.4, v=0.6. Identify the type of object created by the sweep operation.

Solution:

Figure 6.18 shows the ellipse and the intended sweep direction. Without loss of generality, parameter **u** is considered for the basic curve (ellipse) while parameter **v** is along the sweep direction.

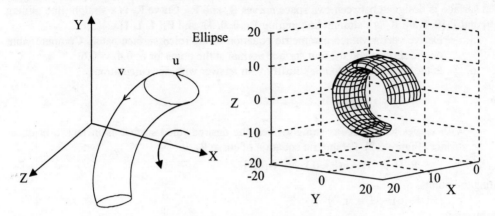

Figure 6.18 Modeling of Elliptic Torus

Vector valued parametric equation of the ellipse is given by
$$P(u) = [x(u)\ y(u)\ 0], \quad \forall\ u \in [0,1]$$

Parameterizing along u,
$$\theta = u.\ \theta_{max}$$
$$= 2\pi u,\ \forall\ u \in [0,1]$$

Thus
$$P(u) = [10 + 3\cos(2\pi u),\ 10 + 2\sin(2\pi u), 0] \qquad \textbf{6.25}$$

Parameterizing along v,
$$\alpha = v.\alpha_{max}$$
$$= \frac{3\pi v}{2}, \forall v \in [0,1]$$

Vector valued parametric equation of the swept surface is given by

$$P(u,v) = [x(u,v),\ y(u,v),\ z(u,v)],\ \forall\ u,v \in [0,1]$$

where,
$$X(u,v) = 10 + 3\cos(2\pi u)$$

$$Y(u,v) = (10 + 2\sin(2\pi u)).\cos\left(\frac{3\pi v}{2}\right)$$

$$Z(u,v) = (10 + 2\sin(2\pi u)).\sin\left(\frac{3\pi v}{2}\right) \qquad \textbf{6.26}$$

Using Equation 6.26, the point on the surface is P(0.4,0.6) = [7.57, -10.62, -6.58]

Figure 6.18 shows the plot of the surface for various values of $u \in [0,1]$. The modeled object is a Torus with elliptic cross section. In practice, the shapes of Donuts, automobile tire (tube) are *Tori*. Any closed curve can be used as the basic curve to create a Torus.

2. A ruled surface is designed between two space curves R_1 and R_2. Curve R_1 is a straight line joining $P_1[1, 0, 0]$ and $P_2[0, 1, 0]$. R_2 is a straight line joining $P_3[0, 0, 1]$ and $P_4[1, 1, 1]$.

     i.   Derive vector valued parametric equation of the ruled surface patch. Compute point on the surface and the unit surface normal at the point for u=0.4, v=0.6.

     ii.   Is the surface *developable*? Justify your answer with due calculations.

Solution:

i)   With curves R_1 and R_2 as straight lines, the desired ruled surface is in fact, a bi-linear surface (Figure 6.7). Parametric equation of curve R_1 is

$$R_1(u) = P_1 + u(P_2 - P_1), \quad \forall u \in [0,1]$$

Substituting the points,

$$R_1(u) = [1\text{-}u, u, 0]$$

In similar manner,

$$R_2(u) = P_3 + u(P_4 - P_3), \quad \forall u \in [0,1]$$

$$= [u, u, 1]$$

Using Equation 6.11, Vector valued parametric equation of the surface is

$$P(u,v) = [x(u,v)\ y(u,v)\ z(u,v)], \quad \forall\ u,v \in [0,1]$$

where,

$$x(u, v) = 1\text{-}u\text{-}v+2uv$$
$$y(u, v) = u$$
$$z(u, v) = v \qquad\qquad \textbf{6.27}$$

Computed point on the surface for **P(0.4, 0.6) = [0.48, 0.4, 0.6]**.

ii)   To check the developability of the surface, Gaussian curvature is to be computed at various points on the surface. Surface equation is

$$P(u) = [1\text{-}u\text{-}v+2uv, u, v], \quad \forall\ u,v \in [0,1]$$

Computing the various partial derivatives,

$$P_u = \frac{\partial\ P(u,v)}{\partial u}$$
$$= [-1+2v, 1, 0]$$

$$P_v = \frac{\partial\ P(u,v)}{\partial u}$$
$$= [-1+2u, 0, 1]$$

$$P_{uu} = [0, 0, 0]$$
$$P_{uv} = [2, 0, 0]$$
$$P_{vv} = [0, 0, 0]$$
$$P_u \times P_v = [1, 1\text{-}2v, 1\text{-}2u]$$

Unit surface normal is

$$N = \frac{P_u \times P_v}{|P_u \times P_v|} \quad \text{for } u=0.4, v=0.6$$

Thus, **N** = [0.962, - 0.19, 0.192]
To compute Gaussian curvature **K**,

$$[A\ B\ C] = [P_u \times P_v] \cdot [P_{uu}\ P_{uv}\ P_{vv}]$$

Thus
$$A = C = 0$$
$$B = 2$$

Gaussian curvature

$$K = \frac{AC - B^2}{|P_u \times P_v|^4}$$

$$= \frac{-4}{[1 + (1\text{-}2v)^2 + (1\text{-}2u)^2]^2}$$

6.28

Figure 6.19 shows the modeled ruled surface (a Hyperboloid) and the variation of Gaussian curvature at different points on the surface. It is seen that Gaussian curvature varies at different points on the surface for $u, v \in [0,1]$ and is always Negative (<0).

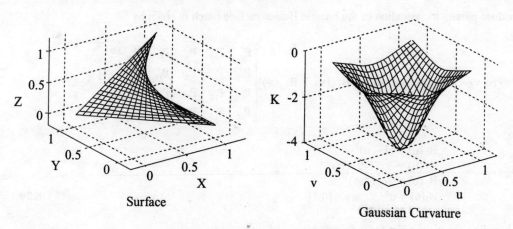

Figure 6.19 Ruled Surface Gaussian Curvature

The surface is thus, *not* developable.

3. A bicubic Bezier surface patch is designed from the set of following control points.

Point	X	Y	Z
P_{00}	0	0	0
P_{10}	0	2	5
P_{20}	0	5	2
P_{30}	0	12	0
P_{01}	2	1	1
P_{11}	2	5	6
P_{21}	2	8	2
P_{31}	2	10	6
P_{02}	4	2	8
P_{12}	4	6	2
P_{22}	4	8	4
P_{32}	4	10	8
P_{03}	6	0	1
P_{13}	6	6	8
P_{23}	6	7	10
P_{33}	6	10	2

Derive vector valued parametric equation of the surface patch. Compute point on the surface at **u=0.5, v=0.6**. Draw the shape of the surface patch.

Solution:

Vector valued parametric equation of the bicubic Bezier surface patch is given by

$$P(u,v) = [B_{0,3}(v) \quad B_{1,3}(v) \quad B_{2,3}(v) \quad B_{3,3}(v)] \begin{bmatrix} P_{00} & P_{10} & P_{20} & P_{30} \\ P_{01} & P_{11} & P_{21} & P_{31} \\ P_{02} & P_{12} & P_{22} & P_{32} \\ P_{03} & P_{13} & P_{23} & P_{33} \end{bmatrix} \begin{bmatrix} B_{0,3}(u) \\ B_{1,3}(u) \\ B_{2,3}(u) \\ B_{3,3}(u) \end{bmatrix}$$

where

$$B_0(u) = (1-u)^3$$
$$B_1(u) = 3u(1-u)^2$$
$$B_2(u) = 3u^2(1-u)$$
$$B_3(u) = u^3 \qquad u,v \in [0,1]$$

6.29

Similar blending functions exist for **v** directions.
Computing blending functions for u =0.5, v = 0.6 and substituting the control points, **P(0.5, 0.6)** is [3.60, 6.30, 4.66].

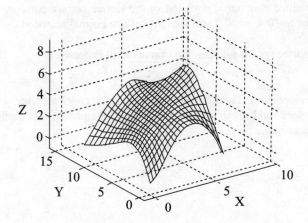

Figure 6.20 Bicubic Bezier Surface

Figure 6.20 shows the plot of the Bezier surface.

6.8 REVIEW QUESTIONS

1. An elliptic cone has its base in X-Y plane and axis along the Z axis. The base ellipse has semi major axis of 5 and semi minor axis of 3. Height of the cone is 5. Derive vector valued parametric equation of the cone. Compute point on the cone surface for u = 0.2, v = 0.3. Is the cone surface developable? Justify your answer with due calculations.

2. A Bezier curve designed from the control points P_0 [0, 0], P_1 [3, 5] P_2 [5,5], and P_3 [5,0] is swept around X axis by 360 degrees. Derive vector valued parametric equation of the swept surface generated. Compute some sample points and plot.

3. A quadratic-cubic Bezier surface patch is designed from the following control point mesh.

Point	X	Y	Z
P00	0	0	0
P10	0	0	5
P20	0	8	0
P01	3	1	1
P11	3	3	8
P21	3	6	2
P02	6	3	3
P22	6	8	6
P02	6	10	2
P03	10	0	0
P13	10	5	5
P23	10	10	10

Derive vector valued parametric equation of the Bezier surface patch. Compute point on the surface patch for u = 0.4, v = 0.7 and the unit surface normal vector at the point.

4. Prove that the above surface patch satisfies the Convex Hull Property.

5. Explain in brief
   i)   All developable surfaces are *Ruled* but the converse is not true.
   ii)   Advantages and limitations of implicit form of equations for surface design compared to the parametric form.

Chapter 7

Geometric Modeling of 3D Objects

We see objects of various shapes, sizes and colour in the world around us every day. Geometric modeling aims at providing a graphical design environment to create such objects, transform and view them in different settings in the 3D virtual world. The geometric models so created are useful for a variety of down line applications such as digital prototyping, CAD/CAM, computer art and animation.

This chapter will present in details, the mathematical basis of the geometric modeling techniques for the synthesis and manipulation of 3D objects.

7.1 ISSUES IN GEOMETRIC MODELING

Geometric modeling forms the core of any CAD/CAM activity as it creates a central repository of product/ part information from where various downstream applications draw upon data. The geometric modeler is thus, required to satisfy many geometric and topological constraints during object modeling to enable design automation. Listed below are some important requirements of a geometric modeler.

- **Versatility (Domain)**
  The modeler should be able to represent a large variety of shapes of solids in the chosen design domain.
- **Validness**
  The representation should not create any *Invalid* or *Impossible* solid which cannot exist in the real world. In particular, the modeler should not represent an object that does not correspond to a solid.
- **Unambiguity**
  The representation should be unambiguous and complete. It should not create doubts nor generate multiple interpretations while reasoning solid models for down line application. This requirement is very important for design automation.
- **Uniqueness**
  The representation should be unique, giving only one way of representing the solid. This is helpful in comparing if two solids are identical.
- **Robustness and Accuracy**
  The representation should be accurate and should always provide valid solid during object transformation and merging.
- **Compactness and Efficiency**
  A good representation should be compact for saving memory. Further it should allow efficient algorithms to compute various properties of objects/ assembly. These requirements are getting more critical for internet based collaborative design situations.

- **User Friendliness**
  The modeler should be user friendly for object creation, manipulation and property computation. It should provide an efficient Query Handling process.
- **Interoperability**
  The modeler should provide good interoperability for seamless integration with the down line tasks. In particular, it should be possible to extract data from the modeler and process it through algorithms written as external programs in C/C^{++}, Java etc.

As of now, there is no universal geometric modeler which can provide all the above characteristics. This is because some of the above requirements are conflicting in nature. Researchers worldwide have been actively carrying out research for the past two decades to develop robust and flexible geometric modeling systems. In the commercial world, geometric modeling systems have been developed for specific part shapes/ application domains around graphical kernels (core) such as ACIS, Parasolids etc. Universal modeling system is however, still a distant dream.

In this book, fundamentals of geometric modeling techniques and their schemas for object representation will be discussed

7.2 TECHNIQUES OF GEOMETRIC MODELING.

A geometric modeler is expected to model a wide range of object shapes from simple planar cuboids to the complex parts in automobile, aerospace or ships. Building universal geometric modeler catering to represent such product range and satisfy stringent modeling requirements (section 7.1) is a very complex, difficult and often impossible task. Geometric modelers are often designed to suit specific part domains and end applications. Three approaches are popularly followed. These are

- Wireframe modeling
- Boundary Representation (B-Rep) modeling
- Constructive Solid Geometry (CSG) modeling

The three approaches follow completely different philosophies for object modeling. Wireframe modeling technique creates the object model as a frame similar to the ones used in the construction of bridges or transmission towers. Boundary representation *(B-rep)* modeling technique constructs a 3D object from its surfaces by *gluing* them to form object edges. The approach is similar to making models of buildings in architecture. Constructive solid geometry *(CSG)* technique follows an approach of constructing 3D object as combinations (addition, subtraction) of some standard predefined solid objects. *Wireframe* and *Boundary Representation* (B-rep) modeling techniques will be discussed at length in the present chapter while the constructive solid Geometry (CSG) technique will be presented in the next chapter (Chapter 8).

7.3 WIREFRAME MODELING TECHNIQUE.

Wireframe model is the simplest of the three approaches to object modeling. It conceives the object as a frame made of a stiff wire bent around the edges of the object. The object model has only *Edges* and *Vertices* and no *Faces*. Topologically the object model is termed as a E-V (*Edge-Vertex*) model.

7.3.1 Wireframe model representation.

A wireframe model has data on the object vertices and constituent edges only. Figure 7.1 shows the wireframe model of a familiar object Cube. The object has 8 vertices [P_1, P_2........P_8] and 12 Edges [E_1, E_2, ...E_{12}]. In the R^3 space, each vertex is represented as a position vector P[x, y, z]

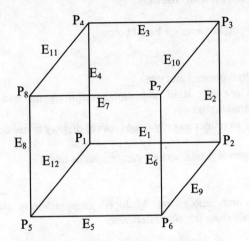

Figure 7.1 Wireframe model of a Cube

Wireframe models use a very simple data structure like array to represent the object geometry and topology. The data structure has two tables viz. Vertex Table and Edge Table. Table 7.1 shows typical data structure for the cube shown in Figure 7.1

Table 7.1
Data Structure for Wireframe model

Vertex Table

Vertex	x	y	z
P_1	x_1	y_1	z_1
P_2	x_2	y_2	z_2
P_3	x_3	y_3	z_3
P_4	x_4	y_4	z_4
P_5	x_5	y_5	z_5
P_6	x_6	y_6	z_6
P_7	x_7	y_7	z_7
P_8	x_8	y_8	z_8

Edge Table

Edge	Vertex - 1	Vertex -2
E_1	P_1	P_2
E_2	P_2	P_3
E_3	P_3	P_4
E_4	P_4	P_1
E_5	P_5	P_6
E_6	P_6	P_7
E_7	P_7	P_8
E_8	P_8	P_5
E_9	P_6	P_2
E_{10}	P_7	P_3
E_{11}	P_8	P_4
E_{12}	P_5	P_1

Vertex table specifies the object geometry while the edge table specifies the topology of the object in terms of connectivity. All the geometric and projection transformations discussed in Chapter 3 and 4 can be applied to the vertex table to transform and view the model. Topology of the object is invariant under the affine and projection transformations.

7.3.2 Characteristics of wireframe models

Wireframe models offer several advantages listed below.

Advantages
- They are simplest to represent and use.
- Wireframe models are very light on memory requirements and can thus, be easily stored, shared and transmitted across net.
- The models are fast to render as only edge (curve) display is needed.

Wireframe models have, however, same serious limitations as under.

Limitations
- Wireframe models are ambiguous. Multiple interpretations about the object shape and geometry are possible from the some database.

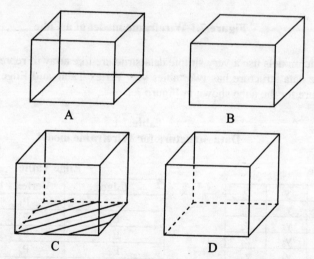

Figure 7.2 Multiple interpretations of Wireframe model

Figure 7.2 shows the wireframe models of four familiar objects. All of them have 8 vertices and 12 edges. Object at A is a cuboidal frame and the one at B is a solid cube. Object at C is a box with open top while the one at D is a box open at both top and bottom. The objects A, B, C, D have respectively zero, 6, 5, and 4 faces. All the 4 objects are valid objects. The wireframe model is ambiguous to specify which one is the intended object.

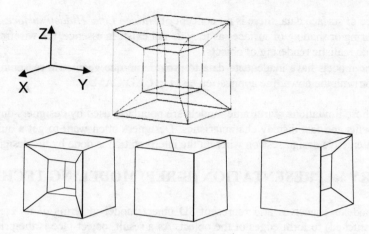

Figure 7.3 Ambiguous Wireframe Model

Figure 7.3 shows the wire frame model of an object A with 16 vertices and 32 edges. It provides 3 possible interpretations (B, C, D) wherein the through slot is oriented along X, Y, or Z direction. These are completely different solids from the design / manufacturing considerations. The wireframe model thus, provides ambiguous representation and is, not suitable for design automation.

- Wireframe models can represent invalid objects. There is no mechanism to validate the topology of objects due to the lack of surface information. Figure 7.4 shows the wireframe representation of objects which are invalid and cannot be produced in real world.

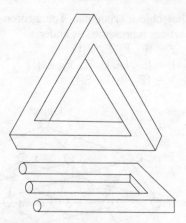

Figure 7.4 Impossible Object

- Wireframe models cannot be used to compute object properties like surface area, volume, moment of interia, weight etc due to the lack of surface information. These properties are important for computer aided design and analysis of objects for stresses, vibration, assembly analysis and manufacturing planning.

- In absence of surface data, there is no concept of *Hidden Line/ Hidden surfaces*. Further there is no coloring or shading of surfaces as they do not exist. In essence, the wireframe models do not provide realistic rendering of objects.
- Wireframe models have inadequate data to provide unique geometric reasoning required for integration with the down line application tasks in CAD/CAM.

Despite these limitations, wireframe models are popularly used by designers due to their light memory requirements and fast display characteristics. Designers often want to get a quick feel of the shape during an interactive design session wherein the interpretation is done by the designer himself.

7.4 BOUNDARY REPRESENTATION (B-REP) MODELING TECHNIQUE.

B-Rep modeler conceives and represent 3D object model in terms of its constituent faces which are glued (stitched) to form edges of the object. As a result, object face is the primary entity in B-Rep modeling. In order that the object model is unambiguous, unique and valid, certain constraints need to be imposed on the object construction and representation. These are discussed at length in the sections to follow.

7.4.1 Object Topology

In B-Rep modeling, object is represented in terms of its faces which may be planar or non-planar like quadrics, ruled or freeform surfaces (chapter 6). Similarly the edges of the object may be straight or made of curves. Topologically however, every object will have constituent Faces (surfaces), Edges (curves), and Vertices (points). Hence for all the future discussions in this book, terms such as Face, Edge and Vertex will be used to denote the object topology.

Figure 7.5 shows a familiar object Triangular Tetrahedron and its associated topology. The object has 4 faces, 6 edges and 4 vertices represented as under

<u>Faces</u>	F = {F$_1$, F$_2$, F$_3$, F$_4$}
<u>Edges</u>	E = {E$_1$, E$_2$, E$_3$, E$_4$, E$_5$, E$_6$}
<u>Vertices</u>	V = {P$_1$, P$_2$, P$_3$, P$_4$}

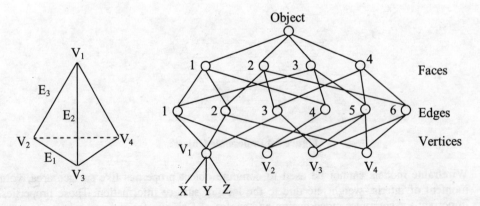

Figure 7.5 Tetrahedron and its Topology

Each vertex can in turn, be represented as a point P[x, y, z]. Figure 7.5 shows the relation between **F**, **E**, and **V** in a hierarchic manner. The nodes in the diagram represent the entities in the set at specific hierarchic level (F,E,V) while the arcs represent the topological connectivity. For example, the face F_1 has 3 edges {E_1, E_2, E_4,} and edge E_1 is incident on 2 vertices {V_1, V_2}. This graph structure indicating topological connectivity between **F, E,** and **V** is very important for representation as well as validation of the object.

7.4.2 Manifold – Vs- Non Manifold objects.

This is an important issue in B-Rep modeling. To understand it, let us look at the topological relationships in the graph shown in the Figure 7.5.

The object (tetrahedron) is regular and symmetric. In a *Top - Down* fashion the relationship is Object ⟶ F ⟶ E ⟶ V while the *Bottom – Up* relation is V ⟶ E ⟶ F
Following observations can be made.

- Each face has 3 edges. (F → E)
- Each Edge is incident on 2 vertices. (E → V)
- Each Edge is shared by 2 Faces. (E → F)
- Each vertex has 3 Edges emanating from it. (V → E)

All objects may not be regular and topologically symmetric like the tetrahedron and so some of the above relations may change. However for an object to be *valid*, some relationship must be satisfied. For example,

- For a face to be closed (valid), it must have Edges ≥ 3.
- For a solid to be closed (water tight), the number of edges from each vertex must be ≥3.

It is important to note that all the valid solids in real world are *Manifold* object. This essentially means that for the object to be solid, number of faces sharing an edge is precisely 2. Such objects are termed as 2- manifold objects. They can be designed and manufactured.

Figure 7.6 shows some non- manifold objects where the number of faces sharing an edge are more than 2. These objects are not feasible in real world. Objects with dangling faces are considered invalid. Researchers are proposing special data structures to handle non – manifold objects and mesh based models. Such cases are out of scope of the present discussion.

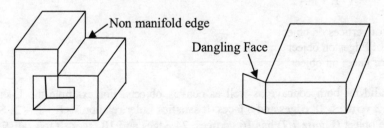

Figure 7.6 Non manifold objects

In this book, geometric modeling of 2- manifold objects will be discussed.

7.4.3 Topology of Polyhedra.

Polyhedron is a solid object whose faces are planar polygons. Object edges are straight lines. It is simple to handle planar polygons from computational view point. As a result, many B- Rep modelers approximate curved surfaces of objects to planar polygons. Polyhedral B-Rep representations are memory efficient but have limited accuracies particularly during intersection algorithms. It is important to study the topology of polyhedra as they provide interesting insight in object modeling.

Polyhedra can be convex or concave based on the shape of the constituent polygons. Figure 7.7 shows some concave polyhedral objects viz a L-shaped and a U shaped object with associated convex and concave polygons. Objects with *Regular* polygons are termed as *Regular Polyhedra*. They have faces having identical number of edges. Cube, Tetrahedron are the familiar examples of regular Polyhedra.

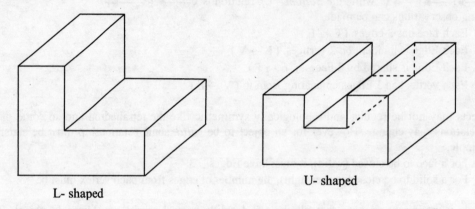

L- shaped

U- shaped

Figure 7.7 Concave Polyhedra

7.4.4 Topological Validity of solids.

Euler had extensively studied the topology of solids and suggested the relation to check their topological validity. If a solid is closed, has valid faces and no tears (holes) in faces,

$$V - E + F = 2 \qquad\qquad 7.1$$

where   V = No. of vertices on object
         E = No. of Edges on object
         F = No. of Faces on object

The relation is valid for both concave as well as convex objects. For example the L-shaped object (Figure 7.7) has 12 vertices, 18 edges and 8 faces. It satisfies Euler relation $V - E + F = 2$. Similarly the U shaped concave object (Figure 7.7) has 16 vertices, 24 edges and 10 faces. It too satisfies the Euler relation $V - E + F = 2$.

7.4.5 Topology of Regular Polyhedra

Regular polyhedra have fascinated mankind since ages. Greeks tried to build several regular shapes and found that only 5 regular polyhydral shapes are possible to be constructed. They considered these objects to have come from heaven and called them as *Platonic solids*.

Why are there only 5 regular polyhedra?

Euler relation provides answer to this interesting question.

A regular polyhedra is a symmetric solid whose faces have identical number of edges. In addition, the number of edges incident on any vertex are identical. Let a regular polyhedra be constructed with topology such that.

- the number of edges incident on any vertex is K_1 and
- the number of edges constituting any face is K_2

The object is a valid solid of the 2- manifold type. Faces of the object do not have any tears (Holes).
For the object to be valid, it must satisfy Euler relation. Thus

$$V- E + F = 2 \qquad\qquad \textbf{7.2}$$

To reduce unknowns, number of vertices (v) and faces (F) can be substituted in terms of edges (E) using the topology constraints K_1 and K_2

$$V. K_1 = 2E$$

and $$F. K_2 = 2E \qquad\qquad \textbf{7.3}$$

Substituting Equation 7.3 in Equation 7.2,

$$\frac{2E}{K_1} - E + \frac{2E}{K_2} = 2$$

Thus $$E = \cfrac{1}{\left[\dfrac{1}{K_1} + \dfrac{1}{K_2} - \dfrac{1}{2} \right]} \qquad\qquad \textbf{7.4}$$

Solution of Equation 7.4 will give the number of edges (E) of the regular polyhedra which is valid. It is difficult to solve Equation 7.4 as it has many unknowns. Some constraints can be put up to solve Equation7.4.

- No. of edges (E) on the object must be an integer
- Edges per face (K_2) and Edges from vertex (K_1) must also be integers.
- For a face to be valid, $K_2 \geq 3$
- For an object to have closed volume, $K_1 \geq 3$.

Solving Equation 7.4 with these constraints shows that only 5 valid regular objects are possible. These are shown in Table 7.2.

Table 7.2
Topology of Regular Polyhedra

No	Polyhedra	K_1	K_2	V	E	F	Figure
1	Tetrahedron	3	3	4	6	4	
2	Cube	3	4	8	12	6	
3	Topaz	4	3	6	12	8	
4	Dodecahedron	3	5	20	30	12	
5	Icosahedron	5	3	12	30	20	

It can seen that object pairs like (Topaz and Cube), (Dodecahedron and Icosahedron) are Duals of each other. Cube and Topaz are also termed as Hexahedron and Octahedron respectively. No other regular polyhedral can be constructed for any combination of K_1 and K_2 other than the five listed in Table 7.2.

Thus there are only 5 *Platonic solids*.

7.4.6 Topology of Non regular Polyhedra

B-Rep modeling technique creates valid solid representations of non regular polyhedra too. The non regular polyhedra will have the following topological characteristics

- Number of edges per face (K_2) may change from face to face on the same object but each face will be a valid (closed space) one.
- Number of edges per vertex (K_1) may also vary including the condition of degeneracy at a vertex.
- The object face may have one or more topological tears (Holes). These tears will lead to inner loops of edges which are distinct from the outer one.

Figure 7.8 shows some typical real life objects having these characteristics. Though these objects are valid, they do not satisfy Euler relation (Equation 7.1) due to the topological variations on the objects.

Poincare modified Euler equation to account for these topological variations to test the validity of B-Rep object models. The modified relation is termed as Euler- Poincare formula.

7.4.7 Euler – Poincare formula

Euler – Poincare formula relates to the topological validity of any solid object which can exist in real world.

For the object to be valid,
$$V - E + F - H + 2P = 2 \tag{7.5}$$

where,   V = No. of vertices on object
E = No. of Edges on object
F = No. of Faces on object
H= No. of Holes (Tears in faces) on object
P = No. of Passeges on object

Two new terms need to be introduced. *Holes* refer to the number of topological tears in faces while a *Passege* refers to a *connected* path on the object through which entry and exit is possible. Typical examples are included here to illustrate the concept.

Figure 7.8A shows a cuboid with a rectangular (blind) cavity/ pocket while Figure 7.8B shows a through Hole in the cuboid.

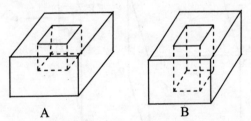

A   B

Figure 7.8 Objects with Hole and Pocket

The object in A has one face (Top) which has a Hole (Tear) shown by an inner edge loop. All other faces do not have such inner loops (Tears). The object thus has, V= 16, E = 24, F = 11, H = 1 and P = 0. It satisfies Euler – Poincare formula for topological validity.

In comparison, the object in B has two faces with holes and one connected passage. Thus it has V=16, E= 24, F= 10, H=2, P=1. It too satisfies Euler – Poincare formula for topological validity of the solid. (Equation 7.5)

7.5 DATA STRUCTURES FOR B- REP GEOMETRIC MODELS.

B-Rep modeling strategy constructs an object in terms of its constituent faces by gluing them along the edges. Data structure for B-Rep is thus, oriented towards efficient representation of the object topology in terms of faces.

Figure 7.9 shows the typical faces of a cube shown in a separated fashion for B-Rep modeling. Each face is considered as the *Oriented Face* in R^3 space. Topologically a face has two sides. In order to model a solid object, it is considered that all the faces are oriented to have their normals projecting to the outside world. In effect, when a user looks at the face of the object , the surface (Face) normal is pointing towards him.

By convention, the face whose normal is projecting outward is considered as *Bright*. For a closed volume solid, all the normals to the inner side of the faces will be pointing at the *inside* of the object. A face essentially comprises of Edges $\{E_1, E_2, ...E_n\}$ and the vertices $\{P_1, P_2.....P_n\}$. To follow the convention of surface normal pointing outwards (Bright face), the vertices (and consequently edges) of the chosen face are ordered in the *counterclockwise* fashion as shown in Figure 7.9.

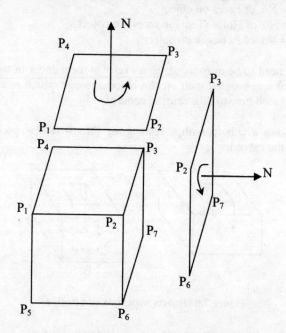

Figure 7.9 Oriented Faces in B-Rep models

Data structure for B-Rep model representation has to satisfy requirements of compact storage, less redundancy and efficient query handling. Quite often a tradeoff is required between these conflicting requirements.

In what follows, three commonly used data structures for B-Rep models will be discussed at length.

7.5.1 Polygon based Data Structure.

Since a B-Rep model is primarily based on boundary polygons, the polygon based data structure looks simple and obvious choice. A polygon mesh is in essence, a collection of faces, edges and vertices which define the shape of the object. The mesh has the capability to represent both manifold and non manifold objects. Polygonal meshes are extensively used for several applications in computer graphics, gaming, computational fluid dynamics etc. Some basic schemas for polygon data structures for B-Rep modeling (mesh representation) will be discussed here. Figure 7.10 shows the face of object (polygon) comprising of vertices and edges.

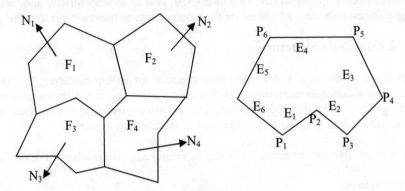

Figure 7.10 Polygon Mesh

The simplest schema to represent polygons could be to have a list of polygons $\{F_1, F_2, F_3,........F_n\}$ wherein each polygon is represented by a list vertex coordinates. Thus

$$F_1 = \{(x_1\ y_1\ z_1), (x_2\ y_2\ z_2),.........\ (x_n\ y_n\ z_n)\}$$

Though simple, it has a limitation that it stores each set of vertex coordinates several times (redundancy). For examples in single precision, each point would need $3 \times 4 = 12$ bits. If an object has 10,000, 4 sided polygons, it would need $10000 \times 4 \times 12 = 480$ Kbytes. This representation is too inefficient and wasteful in memory due to redundancy.

An improvement in representation can be thought of by having a single vertex list like the wireframe modeling (section 7.3). Thus

$$V = \{P_1(x_1\ y_1\ z_1), P_2(x_2\ y_2\ z_2),..........\}$$

The polygon is then an ordered list of vertices i.e

$$F = \{P_1, P_2, P_3 ....P_n\}.$$

In implementation, these will be the pointers to the vertex list V. This schema has the advantage of eliminating the multiple storage of vertices. However it still has limitation is that the shared edges are drawn twice, each with its constituent face. The edge information is *implicit* in data structure which leads to this redrawing.

To remove this problem explicit edge information is sometimes included in the data structure. Thus the polygon is stored as a list of edges in order

$$F_1 = \{ E_1, E_2, E_3 ... E_i, En\}$$

where $E_i = \{V_i, V_j\}$
V_i, V_j represent the vertices belonging to the edge E_i.

The above data structure is efficient for a single polygon or a set of few polygons but become *verbose* as the size grows. Adding explicit information further increases its verbosity. Despite this verbose storage, they need processing to handle even simple queries viz faces sharing an edge etc.

Variety of polygonal mesh data structures exist. Due to their simplicity, they are widely used in different applications (gaming, art) based on the requirements of memory and speed of processing.

7.5.2 Winged Edge Data Structure.

This is one of the most widely used data structure for B-Rep modeling. It was proposed by Baumgart for the efficient representation of B-Rep object topology. It has the capability to represent both manifold and non-manifold objects. Variations in the original winged edge data structure have been proposed such as Half Edge data structures to suit specific requirements.

The winged edge data structure for B-Rep modeling edge is explained in details here.

Data Structure Schema

Figure 7.11 shows a pair of two typical polygons of an object model. The object is considered to be a 2 – manifold one, though this is not the requirement for the data structure. Winged edge data structure focuses upon the representation of the object edge and its associated attributes (data). The data structure in essence, tracks the object edges.

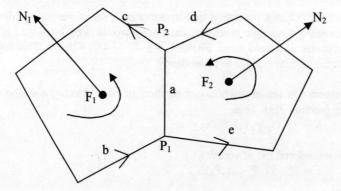

Figure 7.11 Winged Edge Data Structure

As seen in Figure 7.11 the edge **a** is shared by two faces F_1 and F_2 and is incident on vertices P_1 and P_2. The faces in a B-Rep model are *oriented* and so the normal to the faces N_1 and N_2 is projected outwards. This entails that the vertices (and edges) constituting the face are ordered *counterclockwise*. This is shown by the arrows on the faces.

The data structure considers tracking the edge in a specific order. Let the edge **a** be traversed from vertex P_1 to P_2. Then on face F_1, the order will be $P_1 \rightarrow P_2$ while on face F_2 it will be $P_2 \rightarrow P_1$ to keep the counterclockwise notation. Traversing from $P_1 \rightarrow P_2$ on a, the face F_1 will be the *left face* of the edge while the face F_2 will be the *Right face*. Now on each face, the predecessor and successor edges to the edge **a** can be found out. For example on face F_1, the predecessor and successor edges to edge **a** will be **b** and **c**. For face F_2, they will be **d** and **e**.

The winged edge data structure has thus, 8 entries in all as under.

- *Vertices* : Start, End
- *Faces* : Left, Right
- *Edges* : Left Face – Predecessor, Successor
  Right Face – Predecessor, Successor

Table 7.3 shows the Edge table for winged Edge data structure showing various entries.

Table 7.3
Winged Edge Data Structure
Edge Table

Edge	Vertex		Face		Left Face		Right Face	
	Start	End	Left	Right	Predecessor	Successor	Predecessor	Successor
a	P_1	P_2	F_1	F_2	b	c	d	e

The data structure needs the FaceTable and Vertex Table for completeness. These two tables are however, not so exhaustive. Face table lists any one edge belonging to the chosen face. Similarly vertex table lists any edge incident on it. Needless to say that Face Table and Vertex Tables are not unique as multiple choices exist to represent the entries.

Winged edge data structure tries to minimize redundancy of data and enables efficient query processing. Listed below are some typical queries handled efficiently

- Which faces are incident on a chosen vertex?
- Which edges form a chosen face?
- Which edges are incident on a chosen vertex?

An illustrative problem to represent a typical B- Rep model using winged edge data structure is included at the end of this chapter.

7.5.3 STL Data Structure

With the advent of rapid prototyping technology during the past decade, the STL (stereolithography) data format has become very popular for representation of CAD models. The STL format is in essence, a boundary representation (B-Rep) technique wherein the face is tessellated (subdivided) into small triangles called as facets. The size and shape of triangles is decided by the end application as the triangular mesh in effect approximates the exact object surface/ boundary. The triangulation of the surface (face) is *adaptive* in nature which creates triangles of varying geometry and sizes to approximate the curvature of the original surface. Today all the commercial geometric modeling systems have the capability to create and export object CAD models in STL format.STL data format and schema are explained below.

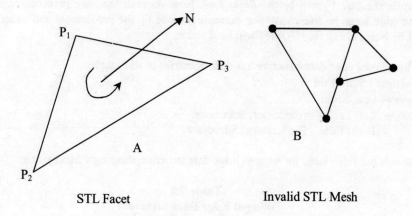

STL Facet   Invalid STL Mesh

Figure 7.12 STL Facet and Data

STL Data Format

A STL file contains list of facets and the associated data. Each facet is identified by a unit normal (N) and the three constituent vertices of the facet. Figure 7.12 A shows a typical facet with its normal N and vertices $[P_1, P_2, P_3]$. In order that the STL facet creates a valid object model, certain constraints need to be satisfied in data format as under.

- Since each facet defines the boundary between the *Outside* and *Inside* of the object, the facet needs to be oriented. This is done in two ways. The facet normal is projecting out of the face. Also the vertices of the face are ordered in counterclockwise direction. It can be seen that the first step is redundant as ordering vertices counterclockwise will automatically give an outward facet normal which can be computed from the vertex data. This redundancy has however, remained in the STL format.

- Each triangle must share two vertices with each of its adjacent triangles. In other words, vertex of one triangle cannot lie on the side of adjacent triangle. Figure 7.12 B illustrates this condition which is a violation of the STL valid format. This is termed as the *Vertex-to-Vertex* rule

STL Data Schema

Each facet data comprises of the normal N $[N_x \ N_y \ N_z]$ and the coordinate data for the three vertices $[V_1, V_2, V_3]$. The facet data will thus, have 12 entries.STL data format can be either ASCII or Binary. ASCII format is readable and hence suitable for testing new software/ CAD interfaces. It is however, too verbose and often impracticable to use. Binary format is, in comparison, very compact and so widely used in CAD model representation, transfer and applications.

The syntax for ASCII, STL file is as under

        Solid   *<name>*

             Facet normal N_x N_y N_z

                    Outer loop

                    Vertex V_{1x} V_{1y} V_{1z}

                    Vertex V_{2x} V_{2y} V_{2z}

                    Vertex V_{3x} V_{3y} V_{3z}

                    end loop

             end facet

        end solid   *<name>*

Bold words are the keywords in the STL format.

7.6 EXAMPLES

1. Using the Euler- Poincare formula, test the topological validity of the objects shown in Figure 7.13 below

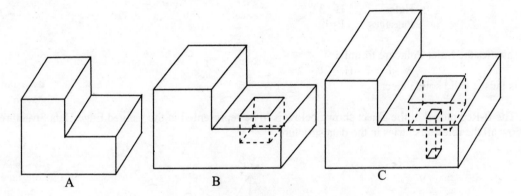

Figure 7.13 Objects with different topology

Solution:

i) The object at A is a concave polyhedron. There are no holes (tears) on any face. The topology of the object is

        Vertices $V= 12$

        Edges $E= 18$

        Faces $F= 8$

        Holes $H= 0$

        Passeges $P=0$

It satisfies the Euler Poincare formula

$$V - E + F - H + 2P = 2$$

It is thus, a valid solid object.

ii) The object at B has a blind pocket. One face of the object, thus, has a hole in it. The topology of the object is

Vertices	V= 20
Edges	E= 30
Faces	F= 13
Holes	H= 1
Passeges	P=0

It satisfies the Euler Poincare formula
$$V - E + F - H + 2P = 2$$
It is thus, a valid solid object.

iii) The object at C has a Blind pocket on one face (similar to B) but there is a through hole at the bottom face of this pocket. The object, thus, has a Passege connecting the outer boundary faces of the object. The topology of the object is.

Vertices	V= 28
Edges	E= 42
Faces	F= 17
Holes	H= 3
Passeges	P=1

It satisfies the Euler Poincare formula
$$V - E + F - H + 2P = 2$$
It is thus, a valid solid object.

2. The B-Rep model of the object shown below is to be represented in the winged Edge Data Structure. Show all the relevant entries in the data structure.

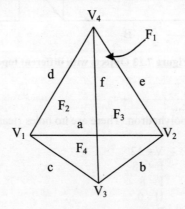

Solution:

The object is a 2-manifold object having 4 vertices $\{V_1, V_2, V_3, V_4\}$, 6 Edges $\{a,b, c, d, e, f\}$ and 4 faces $\{F_1, F_2, F_3, F_4\}$. Using the winged Edge data structure format (section 7.5.2), the relevant topology Tables are designed. These are as under

Edge Table

Edge	Vertex		Face		Left Face		Right Face	
	Start	End	Left	Right	Predecessor	Successor	Predecessor	Successor
a	V_1	V_2	F_1	F_4	e	d	b	c
b	V_2	V_3	F_4	F_3	c	a	f	e
c	V_3	V_1	F_4	F_2	a	b	f	d
d	V_1	V_4	F_1	F_2	a	e	f	c
e	V_2	V_4	F_3	F_1	b	f	d	a
f	V_3	V_4	F_2	F_3	c	d	e	b

Vertex table

Vertex	Edge
V_1	d
V_2	b
V_3	f
V_4	e

Face Table

Face	Edge
F_1	a
F_2	d
F_3	b
F_4	c

3. Geometric model of an object is exported by a CAD software in STL format. The data extracted from the STL file is shown in Table 7.4. Write an algorithm to extract the topology of the object.

Table 7.4
Object Data from STL file

Facet No	Normal			Vertex 1			Vertex 2			Vertex 3		
	x	y	z	x	y	z	x	y	z	x	y	z
1	-1	0	0	0	1	0	0	0	0	0	0	1
2	0	-1	0	0	0	0	1	0	0	0	0	1
3	0	0	-1	0	0	0	0	1	0	1	0	0
4	0.577	0.577	0.577	1	0	0	0	1	0	0	0	1

Solution:

STL file format does not provide explicit information on the edges. Further there is lot of redundancy in the vertex data. Algorithm is designed to address both these problems. Steps of the algorithm are as under.

Algorithm
Begin
1. Create empty sets for Vertex and Edge lists.
2. Choose Facet 1 for processing
3. Transfer 3 vertices to vertex list e.g V= { V_1, V_2, V_3}

4. Identify *Implicit* edges on Facet 1 and transform them to edge list e.g $E = \{E_1, E_2, E_3\}$. Each edge will point to respective vertices in V.
5. Choose next Facet for processing.
6. Identify *New* vertex by comparing coordinates of chosen Facet vertices with those in V.
7. Transfer *New* vertex to V
8. Identify *New* edge of chosen Facet by comparing it with existing edges in E.
9. Transfer *New* edge to E.
10. Repeat steps 5 to 9 till all Facets are processed

End

Processing the given STL file data with the above algorithm, will generate the following object topology.

VertexTable

Vertex	x	y	z
V_1	0	1	0
V_2	0	0	0
V_3	0	0	1
V_4	1	0	0

EdgeTable

Edge	Vertex1	Vertex2
E_1	V_1	V_2
E_2	V_2	V_3
E_3	V_3	V_1
E_4	V_2	V_4
E_5	V_3	V_4
E_6	V_1	V_4

The object is shown in Figure 7.14 with the associated topology.

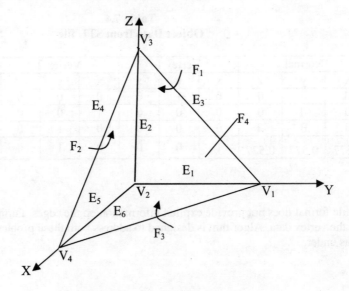

Figure 7.14 Extracted Object

7.7 REVIEW QUESTIONS

1. Bring out the fundamental differences in the modeling philosophies of *Wireframe, B-Rep* and *CSG* geometric modelers.

2. Does the STL data format efficiently represent object topology? Justify your answer.

3. B-Rep model of a cuboidal object of size 5 x 3 x 2 with one vertex at origin is to be represented into STL format. Write down the STL file with relevant entries.

4. Using Euler-Poincare formula, test the topological validity of the following solids.

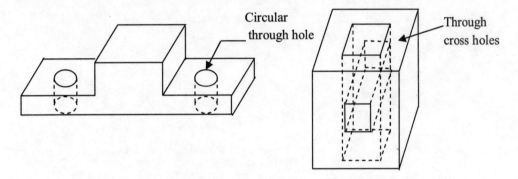

5. Propose an algorithm to process a STL file for the following tacks of Query Processing.
   - To find Facets incident on a chosen vertex
   - To find Facets sharing an Edge
   Explain steps in the algorithm

Chapter 8

Constructive Solid Geometry and Feature based Part Modeling

Geometric models of 3D objects are required for a variety of applications in CAD/CAM, computer graphics and animation. These models need to provide valid and unambiguous representation of solids for automatic interpretation and realistic rendering. Solid modeling using Constructive Solid Geometry (CSG) technique provides such robust object modeling strategy. Feature based modeling is a step further towards interactive part modeling in functional domain.

This chapter presents in details, the mathematical basis of solid modeling using Constructive Solid Geometry (CSG) and Feature based part modeling.

8.1 OVERVIEW OF WIREFRAME AND B-REP MODELING TECHNIQUES

As discussed earlier (Chapter 7) a wireframe model represents object in terms of its Edges and Vertices (E-V) and has no information on object faces. The model is incomplete and often invalid for automatic interpretation. Boundary representation (B-rep) model, in comparison, is informationally complete in terms of object topology (F, E, V) and geometry. It provides unambiguous representation of object which is very suitable for geometric reasoning and automatic interpretation. However the B-rep models have some basic limitations as under

- The object model is constructed in terms of constituent faces which are glued to form object edges. It is *impossible* for a designer to preconceive object faces, construct and glue them to create object model. B-Rep modeling is thus, not suitable for interactive design-synthesis environment.
- B-Rep model gives a very verbose representation of object topology in terms of Faces, Edges, Vertices, Loops, Shells, etc. Though such a representation in terms of low level entities is ideal for extraction of object model data and its further processing for integration with downline application tasks, it is very memory intensive.
- Being a surface based model, B-Rep is very suitable for display but need special algorithms for computing properties like volume, weight, moment of inertia etc.

During mid 80s, Voelcker and Requicha at University of Rochester, USA proposed a completely new paradigm for solid modeling termed as *Constructive Solid Geometry* (CSG). They built a prototype system called PADL (Part Assembly Design Language) to demonstrate the feasibility and efficacy of the CSG concept. Several prototype and commercial solid modeling systems subsequently used the CSG concept to build robust solid modelers. Today CSG has become a key element in any commercial Solid Modeling System.

The Constructive Solid Geometry (CSG) technique will be discussed in details in the sections to follow.

8.2 SOLID MODELING USING CSG TECHNIQUE

The CSG technique conceives the object to be constructed as the combinations (*Addition, Subtraction*) of some standard predefined objects called as *Primitives*. CSG is a Set – Theoretic method of object modeling in which each primitive is defined as a set and the resulting object is derived by operations on these sets. During a typical modeling session, the user will interactively select the desired primitives from the menu of the solid modeler, specify the required parameters and define the set operation to create solid model of the object.

Two essential elements of a CSG based solid modeler are the *Primitives* and the *set operations* on them.

8.2.1 Primitives

CSG solid modeler provides a set of standard predefined solids called as *Primitives*. PADL and its subsequent variants were geared towards the modeling of parts commonly used in mechanical assemblies. They mostly have planar, cylindrical or quadric surfaces. As a result, simple primitives like Block, Cylinder etc are generally provided in the CSG systems. Figure 8.1 shows typical primitives available to the user for design synthesis.

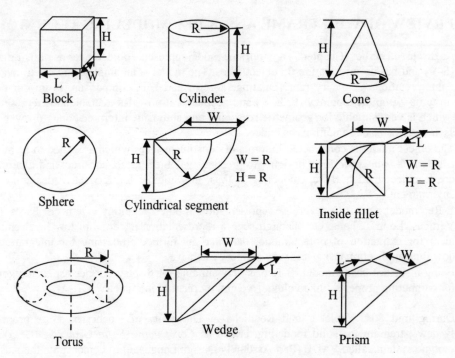

Figure 8.1 Primitives in CSG Solid Modeler

The primitives are internally modeled in terms of surface equations (Half Spaces) and the geometric parameters specific to the primitive. The primitive is modeled in its local coordinate system which is coincident with the modeling World Coordinate System (WCS) in R^3. Figure 8.2 shows a typical cube and cylinder in their respective local coordinate systems and the corresponding geometric parameters. User selects the required primitive and *instantiates* it by inputting the geometric parameters. The solid so obtained is positioned with respect to other objects by carrying out geometric transformations like Rotation, Translation etc (Chapter 3).

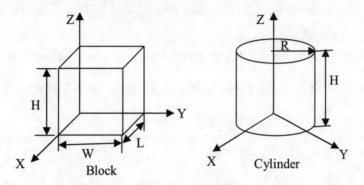

Figure 8.2 Primitives in Local Coordinate System

Though it may appear that too few primitives provided by the CSG modeler (Figure 8.1) will severely restrict its capability in creating variety of shapes it is not so.

8.2.2 Boolean Operations

As mentioned earlier, CSG modeler is a Set Theoretic modeling system. Each primitive is represented as a set and the object to be modeled is considered to be obtained by the combinations of these sets. The set operations are typically *Addition*, *Subtraction*, or finding *Commonality* (Intersection) between the sets. They are termed as Boolean operations. These are in accordance with the laws of Boolean algebra invented by the famous English mathematician George Boole in 18th century. A quick review of set theory and Boolean algebra is presented here.

8.2.3 Set Theory – Overview

A set is a well defined collection of objects which are termed as *members* or *element* of the set. Any quantity or entity can qualify as the member of a set if it lies in the domain of interest. Examples can be the set of natural numbers, all even numbers, all reptiles, all vowels, all mammals, all jacks in a deck etc. In geometry and geometric modeling in particular, set of points is very important. One often tests the location (containment) of a point with reference to the geometric entity, for example whether a point lies on the curves / surface, which side does it lie (inside / outside) etc. A set is generally represented in two ways viz. by *enumeration* or by using Backus-Naur notations. For example using the enumerated notation,

N = {1, 2, 3, 4, 5 ....} is a set of Natural Numbers.
P = {2, 4, 6, 8, 10.......} is a set of even numbers.

The Backus- Naur notation gives a compact notation for the representation of the set in terms of the property or the relation. For example

$P = \{x : x \in 2n \text{ for } n \in N\}$ is a set of even numbers.

$Q = \{x : x \in 2n - 1 \text{ for } n \in N\}$ is a set of odd numbers.

A Universal set **E** is one which contains all the elements of all the sets in a chosen problem domain. In computational geometry, E can be considered to have all 2D or 3D points in space.

A few definitions in set theory are in order
- If the set **A** is a subset of set **B**, then every element in **A** is contained in **B**. It is denoted by $A \subset B$.
- If the set **A** is a *proper* subset of set **B**, then every element in **A** is in **B** and **B** has *at least* one element not in **A**.
- Every set is a subset of itself but not a *proper* subset.
- Complement of a set **A** with reference to the universal set E is the set containing elements in E that are not in **A**. It is denoted by **A'**.

One can form new sets by combining the elements of two or more existing sets. The mathematics which describes this process and laws of combining the sets is termed as *Boolean Algebra*.

George Boole, the famous English mathematician in 18th century invented an algebra for the study of logic which is known as Boolean Algebra. In essence, he attempted to separate the symbols used in mathematical operations from the operands (objects) on which they operate. Treating the operations as independent objects, he was investigating if they could be combined. His work led to Boolean algebra wherein sets are treated as algebraic quantities on which set operations can be done.

Three fundamental set operations were proposed viz, *Union, Intersection* and *Difference*. These are discussed below.

The *Union* operation combines (Adds) two sets **A** and **B** to form a new set **C** which has all the elements contained in **A** and **B**. It is expressed as

$$A \cup B = C \qquad \qquad \textbf{8.1}$$

The *Intersection* operation combines the two sets **A** and **B** to form a new set **C** whose elements are present in both **A** and **B** it is expressed as

$$A \cap B = C \qquad \qquad \textbf{8.2}$$

The *difference* (subtraction) operation between sets **A** and **B** forms a new set **C** which has the elements in **A** but not in **B**. It is expressed as

$$A - B = C \qquad \qquad \textbf{8.3}$$

Venn diagram named after the famous mathematician John Venn gives an excellent tool for the visual representation and interpretation of set theory and the Boolean operations. Figure 8.3 shows the Venn diagram for the *union, Intersection* and *Difference* operations carried out on sets **A** and **B**. For the purpose of illustration, the sets **A**, **B** geometry point are shown as circles, with **E** as the universe. The same representation and interpretation would hold good for sets of other geometry shapes in one, two or three dimensional space. The results of the Boolean operations of *Union, Intersection* and *Difference* are shown in Figure 8.3 for some typical 3D object shapes.

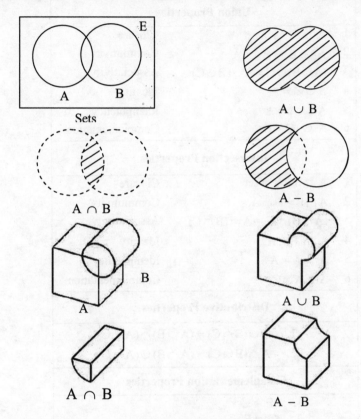

Figure 8.3 Boolean Operations

From the Boolean operations, it can be seen that

$$A \cup B = B \cup A$$
$$A \cap B = B \cap A$$

but $A - B \neq B - A$ 8.4

For example let the sets **A** and **B** be

$A = \{1, 2, 3, 4, 5\}$

$B = \{3, 5, 7, 9\}$

$A \cup B = \{1, 2, 3, 4, 5, 7, 9\}$

$A \cap B = \{3, 5\}$

$A - B = \{1, 2, 4\}, \quad B - A = \emptyset$

Set operations follow the laws of Boolean algebra. They govern the ways by which the sets can be combined to obtain the results. Table 8.1 summarizes important properties of Boolean algebra

TABLE 8.1

Properties of set operations

Union Properties		
1	$A \cup B$ is a set	Closure
2	$A \cup B = B \cup A$	Commutivity
3	$(A \cup B) \cup C = A \cup (B \cup C)$	Associativity
4	$A \cup \varnothing = A$	Identity
5	$A \cup A = A$	Idempotency
6	$A \cup A' = E$	Complementation
Intersection Properties		
1	$A \cap B$ is a set	Closure
2	$A \cap B = B \cap A$	Commutivity
3	$(A \cap B) \cap C = A \cap (B \cap C)$	Associativity
4	$A \cap E = A$	Identity
5	$A \cap A = A$	Idempotency
6	$A \cap A' = \varnothing$	Complementation
Distributive Properties		
1.	$A \cup (B \cap C) = (A \cup B) \cap (A \cup C)$	
2.	$A \cap (B \cup C) = (A \cap B) \cup (A \cap C)$	
Complementation Properties		
1.	$E' = \varnothing$	
2.	$\varnothing' = E$	
3.	$(A')' = A$	
4.	$(A \cup B)' = A' \cap B'$	
5.	$(A \cup B)' = A' \cap B'$	

8.2.4 Object Modeling and CSG Tree

In CSG modeling the user creates object model by selecting primitives in some order and combining them using the Boolean operations. As discussed earlier, the Boolean operations (*Union, Intersection* and *Difference*) are binary in nature; i.e. the operation needs two operands **A** and **B** to create **C**. The CSG modeling is thus, a binary operation.

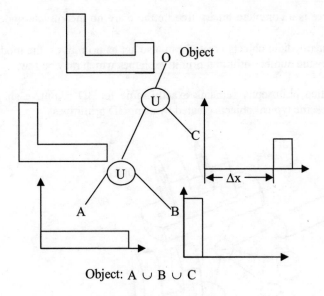

Object: A ∪ B ∪ C

Figure 8.4 CSG Tree

The concept of CSG modeling is explained by taking a simple rectangle 2D primitive. Figure 8.4 shows the steps in the modeling of an object. These are as under

1. User selects primitive **A** and qualifies it with dimensional parameters (Length, width)
2. User selects primitive **B** similarly.
3. Union operation is defined between **A** and **B**
4. User selects primitive **C**, qualifies it with parameters and defines geometric operation (Translation Δx) to position it with respect to **A, B** in the WCS.
5. *Union* operation is defined between **C** and the object created in step 3.
6. Modeling is concluded

Some important observations can be made from Figure 8.4.

- The CSG modeling follows bottoms-up object construction philosophy.
- The transaction history of modeling is captured by the binary tree termed hereafter as CSG tree.
- The CSG tree has 3 types of nodes.
  - The *Root* node at the top of tree is the final object modeled. The *Root* node has no parent.
  - The *Terminal* (bottom) nodes of the tree are called as *Leaf* nodes which are the primitive. They have no child node.
  - The intermediate nodes of the tree have one parent and precisely two siblings (child nodes). Objects at child nodes may be the primitive or an intermediate object generated.
- Boolean operation is stored at each node which operates on its two child nodes.

- The CSG tree is a complete binary tree i.e there are no incomplete nodes except at *the root* and *the leaf*.
- Since the intermediate objects created can also act as primitives, the modeler is not restricted in a sense, by the number of initial primitive shapes which may be few.

The model construction philosophy remains exactly same for 3D primitive shapes. Figure 8.5 and Figure 8.6 shows the some typical objects created by using 3D primitives.

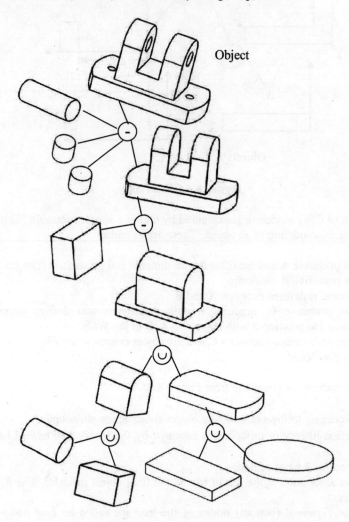

Figure 8.5 CSG Modeling of a Bracket

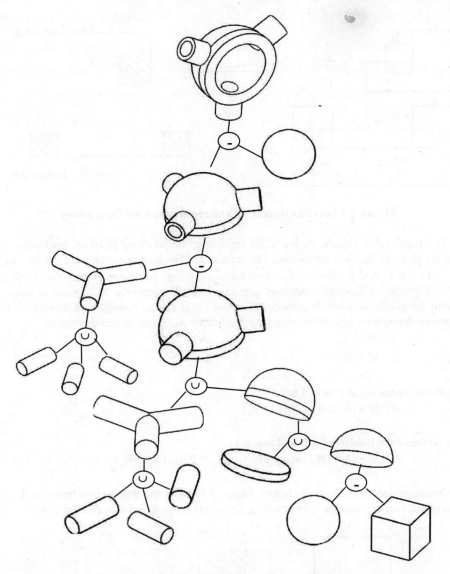

Figure 8.6 CSG Modeling of a Flow Valve Cap

8.2.5 Regularized Boolean Operations

During the CSG modeling, sometimes an invalid object model can get created even though the Boolean operation on the sets may be a valid operation. Figure 8.7 shows an example of intersection between two objects **A** and **B**. The resulting objects **C** generated by the valid operation A ∩ B have an extra dangling edge making it invalid. This is primarily because the dimensionality of the space is not preserved during the set operation. That is the intersections of 2D spaces have resulted into a 1D object. A similar situation may occur for a 3D space too

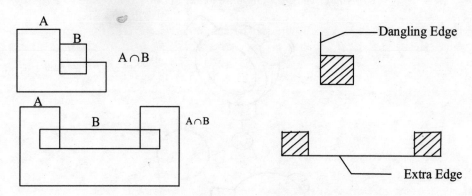

Figure 8.7 Invalid Objects in Non regularized Set Operations

To create valid objects during CSG modeling, *Regularized* Boolean operations are used instead of the standard Boolean operations. The regularized Boolean operations are denoted by *. Thus $A \cup^* B$, $A \cap^* B$, $A -^* B$ denote the *Regularized Union, Intersection* and the *Difference* set operations. Regularized Boolean operators are based on the property of closure of sets. The set representing the primitive/object is considered closed i.e. it has an *Interior* and *Boundary*. Let set **A** has the interior denoted by A_i and the boundary denoted by A_b. It can be represented as

$$A = A_i \cup A_b$$

Similarly $\qquad B = B_i \cup B_b$

The intersection operation $A \cap B$ will be

$$A \cap B = (A_i \cup A_b) \cap (B_i \cup B_b) \qquad\qquad 8.5$$

Using the properties of Boolean algebra (Table 8.1)

$$A \cap B = (A_i \cap B_i) \cup (A_i \cup B_b) \cup (A_b \cap B_i) \cup (A_b \cap B_b) \qquad\qquad 8.6$$

The four bracketed expressions are evaluated. Figure 8.8 shows the intersection between **A** and **B** and the results of the four expressions. The resulting *Union* gives the *Regularized Intersection* $A \cap B$.

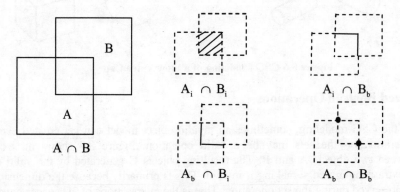

Figure 8.8 Regularized Boolean operation

In any CSG operation, initially the standard Boolean set operations are carried out and the results which violate the dimensionality of space (invalid) are eliminated to get the regularized Boolean operation.

8.2.6 CSG Model Representation

CSG object model is, in essence, a set theoretic procedural model. The transaction history of modeling is represented by the CSG tree (Figure 8.4, Figure 8.5) which is a binary tree pointing to primitives and Boolean operations. For example the object modeled in Figure 8.4 is represented as

$$O = (A \cup B) \cup C$$

CSG modeler stares the procedural model in the form of a compact log file termed as the CSG file.

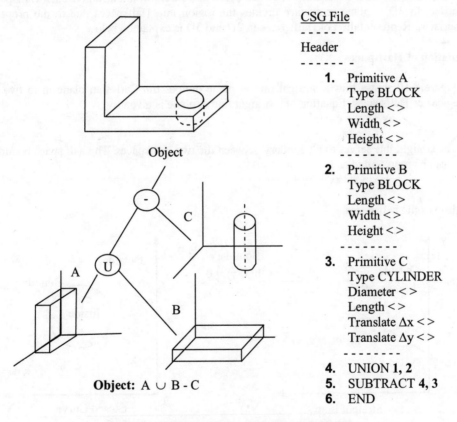

Object

Object: A ∪ B - C

<u>CSG File</u>

- - - - - -

Header

- - - - - -

1. Primitive A
   Type BLOCK
   Length < >
   Width < >
   Height < >

- - - - - - - -

2. Primitive B
   Type BLOCK
   Length < >
   Width < >
   Height < >

- - - - - - - -

3. Primitive C
   Type CYLINDER
   Diameter < >
   Length < >
   Translate Δx < >
   Translate Δy < >

- - - - - - - -

4. UNION 1, 2
5. SUBTRACT 4, 3
6. END

Figure 8.9 Typical Object Model and CSG File

Different modelers use their proprietary formats to store this file. However they essentially capture information on primitives and Boolean operations in the sequence as invoked by the user during the modeling session. Figure 8.9 shows the extract of a typical CSG file. Quite often the CSG file is not accessible to the user but is internally stored by the modeler. It is extremely useful for a variety of tasks in solid modeling such as recreating history in design, redesign, model editing, design history transmission for sharing etc.

8.2.7 Data structures for CSG Modeling

Primitives and solids in CSG modeling are represented by *Halfspaces*. 2D halfspaces are created by straight lines, open and closed curves in plane while the 3D halfspaces are created by planes and surfaces. Combining halfspaces create 2D and 3D finite bounded shapes. In what follows, theory of Halfspaces and representation in 2D/ 3D are presented.

8.2.7.1 Halfspaces – Definition and Representation

In 2D Cartesian space, straight line or open curve divides the Cartesian coordinate plane into two regions called as *Halfspaces*. Without loss of generality, the region of interest on one side can be called as *Inside* and the other as *Outside*. A closed curve like Circle would also give such Halfspaces as Inside/Outside. In 3D, a plane or surface divides the region into Halfspaces due to the property of directed surfaces. Representation of Halfspaces in 2D and 3D is explained below.

Representation of Halfspaces

Figure 8.10 shows how a straight line in plane divides the Cartesian plane in to two semi-infinite regions or Halfspaces. Equation of a straight line in plane is given by

$$ax + by + c = 0 \qquad\qquad\qquad \textbf{8.7}$$

The straight line serves as a boundary between the two Halfspaces. The half space bounded by the line can be represented as

$$h(x, y) = ax + by + c \qquad\qquad\qquad \textbf{8.8}$$

where **h** denotes the halfspaces.

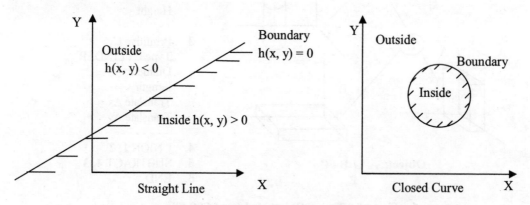

Figure 8.10 Halfspace representation in plane

Boundaries of halfspaces are not restricted to straight lines. Figure 8.10 shows 2D halfspaces defined by closed planar curves such as circle. Any point in the Cartesian region P(x, y) can be classified with respect to the halfspace using equation 8.8 for example.

- If the coordinates of a point P(x, y) yield h(x, y) = 0, then P lies on the boundary of the half space.
- If the coordinates yield h(x,y) > 0, then P lies on the *Inside* half space viz the region of interest.
- If the coordinates yield h(x, y) < 0, then Point P lies on the *Outside* halfspace i.e. outside the region of interest.

In 3D, the halfspaces are represented by Halfspace expression h(x, y, z). The boundary between the halfspaces could be formed by a plane, Quadric surface, an implicit surface, a sphere etc. For example, the halfspace bounded by a plane will be

$$h(x, y, z) = ax + by + cz + d \qquad \qquad \textbf{8.9}$$

As before, a chosen point P(x, y, z) can lie on the boundary, on one side of halfspace or on the other depending upon the evaluation result of the expression in Eqn 8.9. Figure 8.11 shows a typical example showing the boundary, halfspaces and the point classifications.

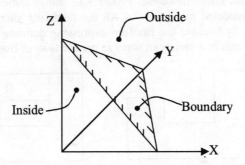

Figure 8.11 Halfspace representation in 3D Space

8.2.7.2 Combining Halfspaces

Halfspaces are combined using the Boolean intersection (∩) operator to create more complex 2D /3D closed shapes. It is very important to maintain the dimensional homogeneity of the space i.e. a 2D halfspace must combine with a 2D halfspace only. In general, if n number of halfspace are needed to create a geometric object O, then

$$O = \bigcap_{i=1}^{n} h_i \qquad \qquad \textbf{8.10}$$

h_i can be a 2D or 3D halfspace. Equation 8.10 can also be written as

$$O = h_1 \cap h_2 \cap h_3 \ldots \ldots \ldots \cap h_n$$

Once two or more finite, valid objects are created, Boolean operations of *Union, Intersection, Difference* can be used to carry out the CSG modeling as discussed in earlier sections. Illustrative problem on halfspace representation is included at the end of this chapter.

8.2.8 CSG Modeling – Advantages and limitations

Solid modeling using CSG technique offers several advantages. Important among them are listed below.

- CSG provides a very user friendly, interactive modeling philosophy which is natural to a designer for design/ synthesis.
- The model representation in terms of CSG file is very compact. It stores only the primitive data and associated Boolean operations (Figure 8.9). In comparison, B-Rep model representation is very verbose in terms of object faces, edges, vertices etc.
- CSG model can be readily used for redesign and object shape editing compared to a B-Rep one.
- Being a set theoretic model, it is robust and produces a valid object model.

CSG modeling has some basic characteristics which sometimes tend to be its major limitations. Important among them are explained below.

- CSG tree is *Non unique*. This essentially means that there are multiple ways of representing the CSG tree to create the same object. Figure 8.12 shows some of the alternative ways of creating the object modeled in Figure 8.4 All the modeling alternatives in Figure 8.12 are valid. This is primarily because the Boolean expressing defining the object in terms of the primitives can be written in a variety of ways as per the laws of Boolean algebra.(Table 8.1)

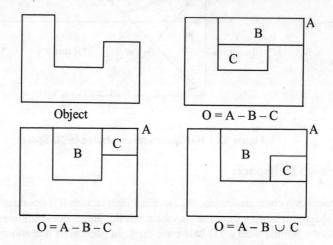

Figure 8.12 Alternative Modeling Stratergies

- The non unique CSG tree representation can, in fact, be an advantage to the designer. The alternative ways provide extreme flexibility to the designer during object model synthesis by not constraining him in anyway. Since the final object modeled is same, it truly captures 'Design intent' during modeling. However the non unique nature of CSG tree is totally unsuitable for automated geometric reasoning of model for linking to downline application task for design automation. As a result, almost all the geometric reasoning algorithms work on B-Rep models despite their verbosity.

- CSG model is very compact in representation being a procedural model. However the model file has no useful data about the object for down line application task, for example it does not give any information on object Faces, Edges, Vertices, Connectivity etc. CSG model is only good for computation of mass, volume, moment of inertia but not surface based properties
- In absence of surface data and topology rendering of CSG models is difficult. Special Ray Tracing algorithms are needed to create realistic rendered images.

In summary, though CSG modeling technique gives a user friendly, flexible modeling strategy, it cannot be used in isolation in CAD/CAM applications

8.3 HYBRID (CSG + B-REP) SOLID MODELERS

Though the B-Rep and CSG techniques provide robust object modeling strategies, they have some inherent limitations which restricts their sole use. CSG technique gives a very user friendly, flexible modeling strategy but the model does not provide any useful topological information suitable for driving a down line application. B-rep strategy, on the other hand, provides a verbose information suitable for application interface but the model construction is extremely clumsy and often impossible. Each strategy has some strengths and weaknesses.

As a result, all the solid modelers used today do not use B-Rep or CSG alone but instead use a hybrid (CSG + B-Rep) strategy for object modeling. (Figure 8.13)The solid modeler has CSG on its front end which provides the user the desired interactively and user friendliness. The CSG object model so created is internally validated and evaluated (tesselated) to represent in the B-Rep format with Faces, Edges, vertices, loops etc. The back end of the modeler provide access to the B-Rep model of the object which can be queried to extract object geometry and topology for developing application programs in CAD/ CAM. Every solid modeler provides this B-rep representation in terms of formats such as IGES, STEP, SAT or similar ones based on the core of the solid modeler. The B-Rep model file provides data on every geometric entity in a verbose manner but is a low level representation of the object model. The user can query the object model using the API (Application Programming Interface) routines provided by the modeler or by using external routines written in C / C^{++}.

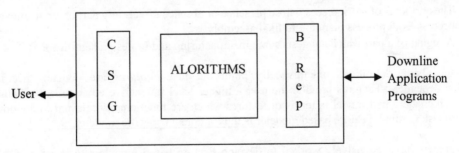

Figure 8.13 Hybrid (CSG + B-Rep) Solid Modeler

The hybrid solid modelers available commercially today thus, keep a dual (CSG + B-Rep) representation of object models internally.

8.4 FEATURE BASED PART·MODELING

8.4.1 Need

Hybrid (CSG + B-Rep) solid modelers provide a robust modeling environment to a designer. It however, needs a designer to conceive the object in terms of the standard primitives provided by the modeler. This is quite often abstract to a user. Engineers and designers do not conceive or think of an object in terms of primitives and the Boolean operations of *Union, Intersection* or *Difference* between them. Instead they look at the object from functional considerations or applications in design, manufacturing, assembly, inspection, or other functional task. Researchers worldwide thus, proposed and implemented the concept of *Feature Based Part Design*. Such modelers are in essence, supersets of Hybrid (CSG + B-Rep) solid modelers as they provide a user friendly, environment to a designer at a higher abstract level.

Feature based part modeling strategy and its related design issues will be presented at length in the sections to follow.

8.4.2 What are Features?

Feature based part modeler enables the user to synthesize the part in terms of functionally important shapes or entities called as *Features*.

What are *Features*?
There is no accepted definition of the term *Feature*. It has been defined by researchers in variety of ways from different application perspectives. Feature is conceived to be

- A region of interest in a part for the purpose of design, manufacturing, analysis or assembly.
- A syntactic means to group data that defines geometric/ functional relation to other elements
- A computer representable data relating to functional requirements, manufacturing processes, or the properties of design
- The attributes of work pieces whose presence or absence affects any part of the manufacturing process from process planning to final assembly.
- A region of a part which suggests some manufacturing and/or design significance.

All the above definitions unequivocally point out that *Features* are domain specific and application dependent. Features provide the user a higher level interactive environment in which the designer can create a part model in terms of features which are functionality relevant to the downline application task in mind. Feature based modeler is thus, a domain specific modeler.

Researchers have extensively worked to develop Feature based modelers to integrate CAD and CAM applications. In particular, majority of work is focused on developing Feature based modelers for machined components and integrating them with Computer Assisted Process Planning (CAPP) systems to link CAD and CAM applications. These systems are developed for Prismatic and Rotational types of parts which are commonly produced on CNC Milling and Turning centers respectively. Of late, Feature based Design and CAPP systems are being developed for sheet metal, die cast, forged and plastic injection molded components.

Though Feature Based Design (FBD) systems provide the designer a functionally relevant interactive modeling system, it constrains a designer to think about the end application right at the start of the design process. This is often cited as one of the major limitations of the FBD system by its critics. The Hybrid (CSG + B-Rep) modeler in comparison does not constrain the user in anyway during the modeling process and thus captures the *Design intent*. Several researchers therefore, have been working to develop geometric reasoning algorithms to understand and extract features from the CAD model data. This line of research is collectively termed as Feature Extraction from CAD models. A number of automated algorithms based on topology Rules, Graphs, Syntactic Pattern Recognition and of late techniques of Artificial Intelligence (AI) like Neural Networks, Genetic Algorithms have been reported in literature. Though several robust algorithms have been reported, in general, they are computationally complex, expensive and often not general enough for a variety of part features. Feature extraction is an ever challenging domain for researchers to develop newer algorithms for intelligent reasoning of CAD models.

Methodology of Feature Based Design (FBD) will be explained, in details, in the sections to follow.

8.4.3 Feature Taxonomy

Feature is the region of interest on a component which suggests some design/ manufacturing activity in the product cycle Figure 8.14 shows some typical prismatic machined part. Looking at it, a machinist will identify that features like *Pocket, Slot, Step, Pattern of Holes* etc need to be machined out from the cuboidal raw stock. Figure 8.15 shows the drawing of a typical rotational part. The machinist will identify that features like *Step, Taper, Groove, Thread, Chamfer* etc are to be machined from the bar stock. The machining process planner uses such feature information for carrying out various activities in *Process Planning* such as Tool Selection, Process Parameters (speed, feed) selection, tool path planning and finally the CNC part program generation. Figure 8.16 shows a sheet metal part with typical features such as *Bend, Hole, Louvre, Dent* etc. The tool designer uses such information for sheet metal process planning and tool design.

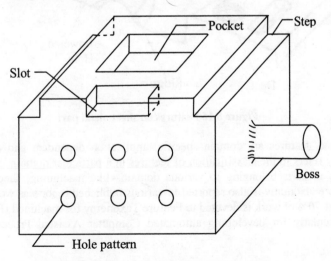

Figure 8.14 Features on Prismatic Machined Parts

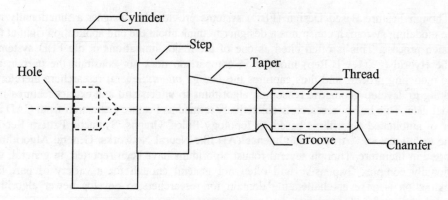

Figure 8.15 Features on Rotational Machined Parts

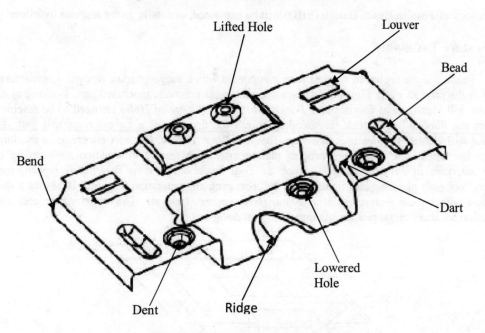

Figure 8.16 Features on sheet metal part

In essence, features are domain specific, application dependent and contextual in nature. Feature Taxonomy refers to the classification of features in a particular domain. NIST has extensively classified features in parts belonging to various domains like machining, sheet metal, dies/ mold, welding etc. Feature taxonomy is also reported in various published reports as well as research papers. It is seen that about 70% of work is devoted to Feature Taxonomy for machined (Prismatic, Rotational) components, particularly for developing automated Computer Assisted Process Planning (CAPP) systems.

To explain the concept of Feature based modeling (FBM), the taxonomy of prismatic machined part features will be used in this chapter.

In general, features can be classified according to various viewpoints as under

- Design/ Manufacturing features
  This is based on intended use of features such as *Shape design, Analysis, Assembly, Machining, Inspection* etc.
- Protrusion/ Depression features
  Features which protrude from base surface such as a Boss are termed as *Protrusion* features while those which need to be carved out from stock are the *Depression ones*. All machining features like Pocket, Slot, Hole etc are Depression features Figure 8.14 shows various Protrusion and Depression Features.
- Face Based / Edge Based features
  Features such as Pocket, Holes lie completely within the face while features like Step, Slot cause variations in the boundary (edge) of the face. The former are termed as Face based while the latter are termed as Edge based (Figure 8.14)
- Special features
  These include connective features like Blend, Fillet, and Chamfer between the faces. Compound features, Intersecting features, Feature in Features etc are the special cases of Features depending upon the design domain.

In feature based modeling, user is provided with a library of features. The modeling essentially follows the steps as under

- Feature Selection
- Feature Qualification
- Feature Model Validation
- Model representation and storage

Author has designed and implemented a Feature based modeler (FBMod) for prismatic machined components commonly produced on 3 axis CNC machining centers. It is a part of an Internet based Process Planning System (WebNC) indigenously developed by the author at the Computer Aided Manufacturing Lab, Mechanical Engineering Department, Indian Institute of Technology , Bombay, India.(http://webnc.cam.iitb.ac.in)

Various steps in feature modeling are explained with reference to FBMod.

8.4.4 Feature Selection

FBMod targets on prismatic machined parts such as gearbox housing, brackets, frames etc commonly produced on 3 axis CNC machining centers. Extensive study of actual components in shop was carried out to identify feature taxonomy. In addition relevant research literature was referred.

FEATURE TAXONOMY

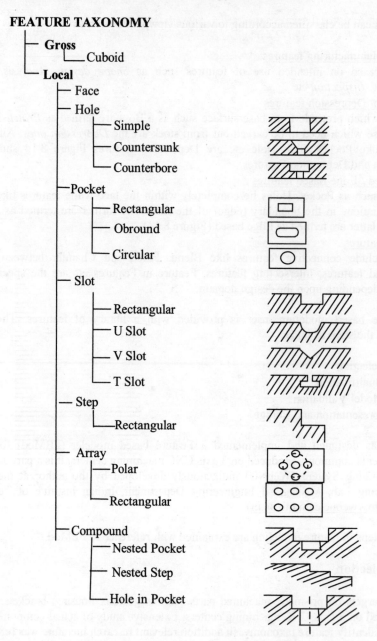

Figure 8.17 Feature Taxonomy in FBMod.

Figure 8.17 shows the Feature Taxonomy used in FBMod. The part model is conceived to have two types of features viz the *Global* feature and the *Local* features. The global feature is a cuboidal shape upon which the designer can interactively position the Local features. Global shape can be created by specifying the cuboid parameters or by using the sweep operations (Chapter 6).

For modeling the local features, six generic families of feature shapes have been identified. These are *Face, Hole, Pocket, Slot, Step and Pattern* (Array). Each feature family can be further divided into subfamilies. For example

<u>Hole</u>: Simple Hole
   Counter Bore
   Counter Sunk
   Threaded

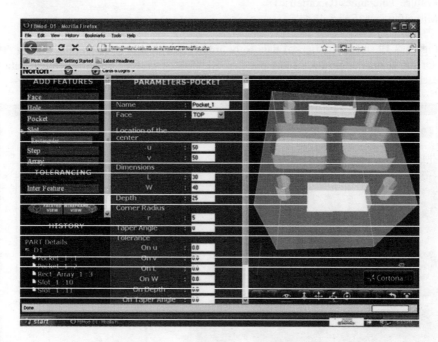

Figure 8.18 Design screen of Feature based modeler – FBMod

Figure 8.18 shows the menu screen of FBMod. User can pick features from the left pane of screen; input its parameters in the middle pane with reference to the Figures in the rightmost pane. History Tree at the left bottom of the screen captures the Transaction History of modeling. It lists the features selected by the user in order. The History Tree can be used for Feature Selection, Editing, and even Deletion.

8.4.5 Feature Qualification

Feature qualification refers to the input of the parametric data of each feature during the modeling phase.

Feature based modeling essentially enables a user to conceptually design, position and orient the chosen feature on the face of the gross shape (stock) in an interactive manner. Feature data is inputted by the user in the design templates which are the forms having data fields specific to each feature. The feature data comprises of the following functional data fields.

- Feature Type < >
- Feature Identification No. < >
- Feature Access Face < >
- Feature Geometry Parameters < >
- Feature Location Parameters < >
- Feature Orientation Parameters < >
- Feature Tolerances (Dimensional) < >
- Feature Tolerances (Geometric) < >
- Miscellaneous (Special) < >

Figure 8.19 shows the typical data to be inputted to design a rectangular pocket. The design template for the feature is in the middle pane of the FBMod screen. Specific feature data templates exist for each feature type.

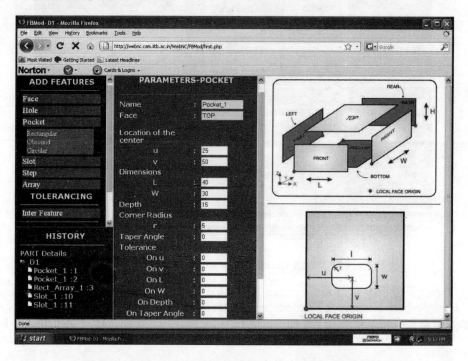

Figure 8.19 Feature based Modeling of a rectangular pocket

8.4.6 Feature Validation

Feature Validation is an important phase in the Feature based modeling. Since the modeler creates components which are domain specific, both geometric as well as functional validation need to be automatically done by the system. For example, a feature based modeler for machined parts must carry out the validation checks to see that the part is topologically valid as well as machinable.

In FBMod, two types of validation checks have been implemented viz the *Syntactic* and *Semantic* checks. These are explained below

Syntactic Checks

Syntactic check refers to the validity of the data which forms the parametric input to each feature. They essentially ensure the correctness of the data type. Typical syntactic rules can be as under
- No data field is empty
- Dimensional data of a feature must be a numeric data
- No feature dimension can be negative
- Dimension data cannot have a special character like - @, $, etc
- Feature name can be Alphanumeric with special characters

Semantic Checks

Semantic checks are essentially meant to validate the model to ensure that it is a valid design in real life and can be produced. These checks are termed as Design for Manufacturing (DFM) checks. Incorporation of such checks right at the stage of modeling is a very important step toward *Intelligent Product Modeling*. It will significantly compresses the lead time from design to manufacture by avoiding further errors in manufacturing, rework and associated design revisions.

In FBMod, feature specific semantic checks have been designed and implemented. Few typical validation rules are as under.
- Face based feature (Pocket, Hole, Array) must lie completely within the face boundary.
- The feature boundary must not intersect or be tangential to the face boundary.
- Enough material should be there between features and face boundaries from strength as well as manufacturing considerations. This will be based on part material, machining process capability and finally the shop practices.
- Radii of fillets should be realistic from manufacturing considerations (tool sizes).
- Feature boundaries should not intersect with each other unless it is intended by the designer. A warning alert should caution the designer to review the feature data before proceeding further in modeling.

8.4.7 Feature Model Representation

During the modeling session, user selects features one by one, and qualifies them by inputting parameter data. The modeling system automatically validates the feature data using the validity checks discussed earlier. Transaction history of modeling is captured as History Tree shown in Figure 8.18.

The feature based model (FBMod) internally represents the feature data in the form of an ASCII feature file. The feature based modeler uses this data file to create solid model of the object using Boolean Operations between the block (Gross shape) and the local feature. The model is finally stored in the B-rep format. Thus internally the feature based modeler maintains object representation in terms of Feature file (*.prt) as well as the B-Rep structure. Figure 8.20 shows the feature file for the part having a rectangular pocket (Figure 8.19). The format of this file used in FBMod is native but similar to the STEP standard. Various modelers may use their own variations of this format.

```
PART_ID 1
MATERIAL Aluminium
UNITS mm
FEATURE Gross
Type Extrusion
SubType Block
ID 0
Name Block
Reference WCS
Location 0.000000 0.000000 0.000000
SubType Block
Orientation X 1.000000 0.000000 0.000000 Y 0.000000 1.000000 0.000000 Z 0.000000 0.000000
1.000000
INFORMATION
FeatureID 0
Dimension:Length:Width:height
Value:100;100;100
ACCESS_DIRECTIONS:-
1.000000,0.000000,0.000000;0.000000,1.000000,0.000000;1.000000,0.000000,0.000000;0.00000
0,-1.000000,0.000000;0.000000,0.000000,-1.000000;0.000000,0.000000,1.000000
END INFORMATION
END FEATURE
FEATURE Local
Type Pocket
SubType Contour
ID 1
Name RectangularPocket
Reference FEATURE_ID 0 FEATURE_FACE_ID 5
Location 50 50 100
Orientation X 1.000000 0.000000 0.000000 Y 0.000000 1.000000 0.000000 Z 0.000000 0.000000
1.000000
INFORMATION
FeatureID 1
Dimension:Length;Width;Depth;Cornerrad;Taper
Value:60;45;12;5;0
ACCESS_DIRECTIONS:0.000000,0.000000,1.000000
END INFORMATION
END FEATURE
END PART
```

Figure 8.20 A Typical Feature Based Part file in FBMod

The feature data is internally represented in the modeler using Object Oriented Programming System (OOPS) concept. Main advantage of using OOPS structure is its modularity and extendability. The feature data fields can be extended if the need arises later to add new features.

8.4.8 Feature Mapping.

One of the main advantages of using OOPS structure for Feature representation is its ability to associate Procedures with the feature data. This is very important to create modularity in software development. It can be considered to be *Feature mapping* in a sense. For example, the data of the feature (pocket) shown in Figure 8.19, Figure 8.20 can be mapped to a variety of activities in Computer Assisted Process Planning (CAPP). Typical among them could be

- Selection of cutting tool (Size, Type)
- Selection of machining process (milling) and strategy of machining
- Selection of process parameters for machining (Speed, Feed)
- Tool path planning (Zigzag, Spiral)
- CNC part program generation and post processing

To automatically carry out these tasks in CAPP, suitable procedures can be embedded (associated) with the OOPS feature data structure. A similar strategy would exist for other downline application tasks which would use the feature data to develop application programs in CAM. Feature data in OOPS format, provides seamless integration between CAD and CAM.

8.4.9 Advantages of Feature based Modeler

Feature based modeling strategy provides several advantages compared to the Hybrid (CSG + B-Rep) modelers. Important among them are summarized below.

- Feature based modeler (FBM) provides a very flexible, interactive design environment familiar to the user in terms of the intended downline application domain.
- It provides a robust and efficient system for a specific application and thus enhances *Design Productivity.*
- It provides *Quick Turnaround* in design, particularly during Prototype development and product redesign phase. It is very suitable for variant design.
- It embeds a robust validation mechanism based on *Design for Manufacturing* (DFM) concepts. Requirements and constraints of downstream activities are available to the designer right at the modeling stage.
- Feature based model file is very light on memory and can be shared/ transmitted easily for interoperability.
- OOPS data structure used in feature representation enables integration with downline application software. It thus, enables seamless integration between CAD and CAM.

Today Features Technology has become a de facto standard in Product Modeling.

8.5 VARIATIONAL GEOMETRIC MODELING.

Variational modelers are essentially Constraint based geometric modeling systems which enable the designer to create variant designs by tweaking/ deforming a model. The objective is to further increase the interactivity and user friendliness during geometric modeling. Feature based modelers are the ideal forms of representations to incorporate constraints during design mainly due to their OOPS data structure.

Researchers and CAD developers worldwide are actively working in the area of developing variational modelers. Variety of techniques are being experimented to incorporate constraints in the models. In what follows, an attempt has been made to present some basics of constraint based variational modeling.

8.5.1 The Constraints

Constraints incorporated on geometric models primarily fall in two categories viz *Geometric* constraints and the *Functional* constraints.

Geometric constraints refer to the dimension and attitude (orientation) of the part features. Functional constraints relate to the part function and are often overall goals for the Optimum design of part/ assembly, for example optimize the shape of a part for maximizing strength/ weight ratio. Functional constraints are often complex problems involving CAD geometry modeling, analysis (FEM/CFD) and multiparameter optimization techniques. These are still in infancy even in research. They are thus, out of scope of the present book.

Typical examples of geometric constraints in CAD modelers are discussed below.

8.5.1.1 Dimensional Constraints

Quite often, the dimensions of an object/ part are related to some key dimensions of its feature. Figure 8.21 shows the typical example of a fastener bolt. It can be seen that variants of this shape can be obtained by inputting key dimension D. This in turn, designs a family of parts having similar features (shape) but with varying dimensions. Such parts are commonly required in creating standard part libraries used in mechanical design of assemblies, Jigs and Fixtures, Fasteners. Hydraulic circuits etc. Many CAD modelers provide utilities to create such variant designs of standard parts in terms of some key dimensions.

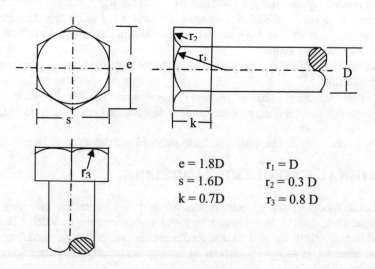

$e = 1.8D$ $r_1 = D$
$s = 1.6D$ $r_2 = 0.3\ D$
$k = 0.7D$ $r_3 = 0.8\ D$

Figure 8.21 Parametric Design of Hexagon Headed bolt

Dimensional constraints are, in essence, the proportions (relations) between various governing parameters (dimensions). These could be predecided or user specified in the modeling system. Knowing these constraints, the proportions of other elements (dimensions) are computed by the modeler and the variant design is carried out. Today many Computer Aided Drafting systems, enable user to interactively input geometry constraints for 2D shapes. Same concept is being extended to 3D object modeling using Feature modeler.

8.5.1.2 Attitude Constraints

These essentially relate to the *Orientation* and *Location* constraints between features. Commonly used attitudinal constraints include *Parallelism, Perpendicularity, Coaxiality, Concentricity, and Symmetry* to name a few.

If a CAD system permits the user to input such constraints during modeling, the model editing phase has to maintain these to ensure validity of the modeling operation. Figure 8.22 shows a simple example in 2D modeling to explain the concept.

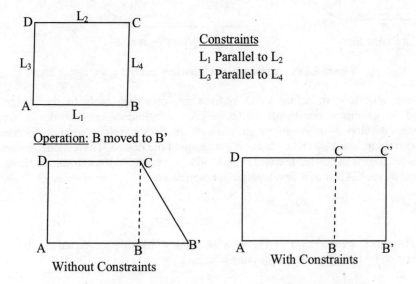

Figure 8.22 Constraint Based Modeling in 2D

The shape of the object is edited by dragging one vertex. Figure 8.22 shows the resulting shape *with* and *without* constraints. Figure 8.23 shows an object where co-axiality constraint is put between the two holes. The show resulting shapes in model editing with and without constraint are also shown.

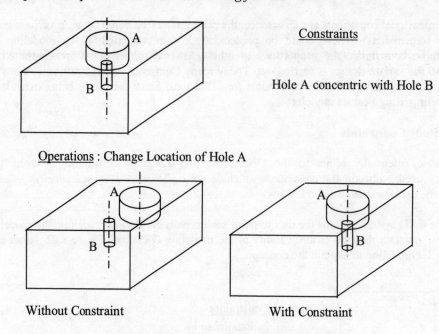

Constraints

Hole A concentric with Hole B

Operations : Change Location of Hole A

Without Constraint

With Constraint

Figure 8.23 Constraint based Modeling in 3D

Today many geometric modeling CAD systems are providing tools to the designer to incorporate 2D and 3D geometric constraints of dimension and attitude during the design process. Design rules are being developed for incorporation into CAD modeling system based on specific end applications such as forging, sheet metal or plastic injection mold/die design. These modeling systems are being termed as Intelligent (Knowledge based) CAD modeling systems. Variational modeling is an area actively pursued by the CAD model developers and researchers.

8.6 EXAMPLES

1. Represent the 2D object space shown in Figure 8.24 using the Halfspace equations. Check the validity of representation by taking some sample points in space.

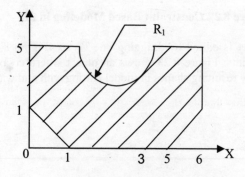

Figure 8.24 2D Object Space

Solution:

The 2D object in R^2 space Figure 8.24 can be obtained as the intersection of geometric entities (1 to 7) shown in Figure 8.25

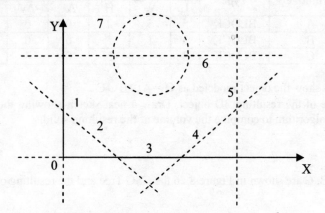

Figure 8.25 Halfspace Representation

Entities 1 to 6 are infinitely long straight lines which divide the region in two halfspaces. Entity 7 is a closed one (Circle) which divides the region as inside/ outside. By convention, the region of interest on the object is taken as positive. Writing the Halfspace equations for the entities, one gets.

$$h_1(x_1y) = x$$
$$h_2(x_1y) = x + y - 1$$
$$h_3(x_1y) = y$$
$$h_4(x_1y) = 5 - x + y$$
$$h_5(x_1y) = 6 - x$$
$$h_6(x_1y) = 5 - y$$
$$h_7(x_1y) = (x-3)^2 + (y-5)^2 - 1 \qquad \textbf{8.11}$$

The 2D object space shown in (Figure 8.22) is the intersection of the above Halfspace equations. Thus

$$2D\ Space = \bigcap_{i=1}^{7} h_i(x,y) \qquad \textbf{8.12}$$

To check the validity of Equations 8.11 and 8.12, choose points P(0,0), P(2,2)
It is seen that P(0,0) gives negative result for one of the halfspace equations in Equation 8.1. This indicates that it lies outside the region of interest P(2, 2) gives positive result for all the halfspace equations in Equation 8.11 indicating that the point lies in the region of interest of the object.

2. A CSG modeling file is processed to extract the history of transaction during modeling. Data of the primitives used in modeling is listed below

Primitive	Type	Parameters			Transformations		
		L	W	H	Δx	Δy	Δz
A	BLOCK	5	5	2	0	0	0
B	BLOCK	3	1	5	0	0	0
C	BLOCK	3	1	5	0	4	0

Boolean operations show the object modeled as $O = A \cup B \cup C$.

Compute the shape of the resulting 3D object. Draw a neat sketch showing the primitives and final object. Propose an algorithm to compute the volume of the resulting solid.

Solution:

The primitives A, B, C are shown in Figure 8.26 the CSG Tree and the resulting object is shown in Figure 8.27

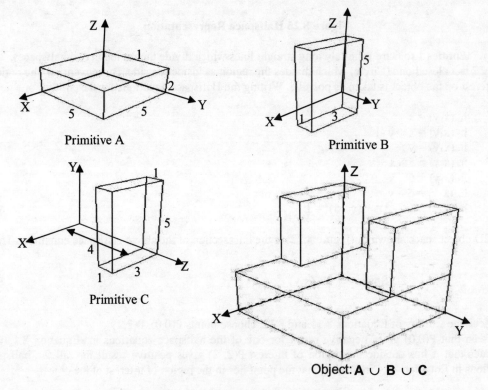

Primitive A

Primitive B

Primitive C

Object: A ∪ B ∪ C

Figure 8.26 CSG Modeling Operation

It is seen that the resulting object is the union of the primitives A, B and C. Figure 8.27 shows the Venn diagram indicating the Boolean operation O = A∪B∪C.

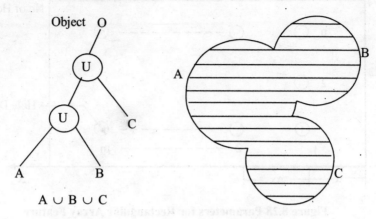

Figure 8.27 CSG Tree and Boolean Operations

Volume of object O can be computed as under

Volume (**O**) = Volume (**A**) + Volume (**B**) + Volume (**C**) - Volume (A∩B) - Volume (A∩C)

8.12

From given data, the volumes can be computed. (Figure 8.24)

Volume (**A**) = 50
Volume (**B**) = 15
Volume (**C**) = 15
Volume (A∩B) = 6
Volume (A∩C) = 6

Using Eqn 8.12, the volume of object **O** is

Volume (**O**) = 68 cubic units

3. In a feature based modeler for machined parts, a new feature to model a rectangular pattern of holes is to be incorporated. Design the feature pattern and its associated parametric data fields.

Solution:

The feature is generic in nature and should cater to the family of rectangular pattern of holes. To incorporate variant design possibility, feature pattern variables are designed. Designer will input data for these data fields (variables) during the modeling session. Figure 8.28 shows the generic sketch of the feature pattern

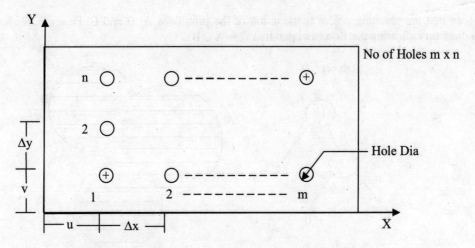

Figure 8.28 Parameters for Rectangular Array Feature

Listed below are the data fields identified.

- Feature Type < >
- Feature ID   < >
- Feature location Face   < >
- No. of Holes – X Direction (m) < >
- No. of Holes – Y Direction (n) < >
- First Hole – X Coordinate (u) < >
- First Hole – Y Coordinate (v) < >
- Pitch – X Direction (Δx) < >
- Pitch – Y Direction (Δy) < >
- Hole Diameter (D) < >
- Hole Depth (H) < >

Figure 8.27 shows the OOPS data structure for the feature.

8.7 REVIEW QUESTIONS

1. For sets **A** and **B,** prove that $\mathbf{A - (A - B)} = A \cap B$

2. In a CSG modeler, a sphere (A) is created with center at origin and radius 5. Three cylinders (B,C,D) are created along X, Y, Z axes respectively each of radius 1. Boolean operations of **O** = **A – B – C – D** is carried out to create the final object. Represent the final object using Halfspace equations. Verify the validity of the 3D space in the object by checking for some sample points.

3. For CSG modeling of the object shown in Figure 8.26(Problem 2), it is proposed to use Boolean subtraction (-) operator only. Draw the CSG tree. Compute the sizes of the primitives to create the object.

4. It is desired to add a new feature *Polar Array Pattern* to a Feature based modeler for machined part. Identify suitable parametric data fields to design this generic feature. Represent using OOPS data structure. Draw a neat sketch.

5. Compare critically the characteristics of Hybrid (CSG + B-Rep), feature based modeler and variational geometric modelers. Bring out the advantages and limitations of each approach.

Chapter 9

Product Data Exchange Standards

Product data forms an important repository which drives variety of technical and business activities of the product life cycle. These typically include product design, analysis, production planning and control, manufacturing, assembly, sales, marketing and finance. Representation of product data is thus, extremely important to seamlessly integrate these product life cycle activities to enable efficient product development and other business functions.

This chapter will present in details, the Product Data Exchange Standards for integrating CAD/CAM activities for Product Lifecycle Management (PLM).

9.1 PRODUCT DATA AND INTEROPERABILITY

Product development activities in an organization can be broadly classified into two categories viz. *Technical* and *Business*. Product data has accordingly two layers, *Geometric* and *Non Geometric* which cater to the specific requirements. Geometric data targets upon the design and manufacturing activities of product development. These essentially form the core activities in CAD / CAM viz Geometric / Solid modeling, Drafting, Product Function Analysis (Stress, Kinematics, Assembly), Visualization, CNC programming, Digital Prototyping / Simulation to name a few. Geometric data governs the product shape, size, topology and engineering product specifications for design and manufacturing activities.

Non geometric product data, on the other hand, refers to the Production Planning and Factory Management. In particular, it refers to the production volume, capacity planning, scheduling, inventory management, materials, sales, marketing and financial data. It is distinct from the geometric data particularly because the shape, size and dimensions of the product are not that important to govern the functions of factory management.

Due to the differing nature and requirements of geometric and non geometric data, separate database structures have been developed in the CAD/ CAM / CIM environment. For example, the relational Data Base Management system (RDBMS) was considered to be the most suitable DB system for the representation of non geometric data.

Since the past decade, comprehensive product data standard is being developed internationally to unify the technical and business data from PLM considerations.

9.2 EVOLUTION OF DATA STANDARDS – HISTORICAL PERSPECTIVE.

Introduction of Computer Aided Design and Drafting (CADD) tools in the early 70s provided very significant benefits in terms of increase in productivity, flexibility and the ease of revision compared to the traditional paper based designs/ drawings. However by early 80s problems started on the inability of sharing data across different CADD systems. Each CAD developer was using his own native data format for representation of the CAD data (model, drawings) which could not be exchanged / integrated with the other CAD system. This posed a serious threat to the basic concept of integration of design/ manufacturing activities to create an integrated CAD/ CAM environment.

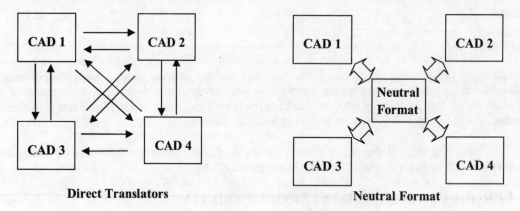

Direct Translators   **Neutral Format**

Figure 9.1 Techniques for Product Data Exchange

A partial solution to this problem was proposed by some CAD developers in the form of dedicated *Data Translators*. The data translators would read in the CADD data in one format and convert it to the format of the other software. Figure 9.1 shows conceptually the data flow diagram for the direct data translators. It seen that for n distinct CAD systems, number of direct dedicated translators needed would be n (n-1). Developers soon realized that this would be infeasible, particularly when n becomes large.

It was thus, increasingly felt that if a common Neutral data format acceptable to CAD developers is agreed upon, the requirement for data translators would get drastically reduced to 2n for n distinct CAD systems (Figure 9.1). This was the genesis for evolving the first data exchange standard called as Initial Graphics Exchange Specification (IGES).

9.3 INITIAL GRAPHICS EXCHANGE SPECIFICATION (IGES)

The idea of having a neutral file format for the exchange of CAD data nucleated in September 1979 during an industry meet called as Integrated Computer Aided Manufacturing (ICAM) in USA. It was attended by CAD developers like Computer Vision, Applicon, users like Boeing, General Electric, Xerox, Air Force, Navy, NASA and the representatives from National Bureau of Standards (now known as NIST). Apprehensions of CAD vendors in sharing their private data formats were soon dispelled. It was unanimously agreed to initiate a project termed as IGES for evolving neutral data file format for exchange of CADD data.

In formulating the goals of this new CAD translation project, a *minimalist* approach was followed. IGES was meant to mean the following

I – *Initial (interim)*, not to replace work of ANSI standard

G – *Graphics,* not to address complete geometry

E –*Exchange,* not to dictate developers who are free to implement their own internal databases.

S – *Specification*, not imposing as a standard.

First version of IGES was adopted as the ANSI standard Y 14. 26 M – 1981 by NBS. Subsequently many revisions of IGES were carried out from time to time. With each revision, the scope of IGES was expanded to include more geometric entities of the product models. The last published revision was the version 5.3 in 1996.

Similar to IGES, parallel efforts to evolve neutral data file exchange formats were attempted which resulted in standards like VDA in Germany. By 1994, an effort to formulate an International standard known as STEP (ISO 10303) started. As of now STEP has been significantly developed for various application domains. The development of STEP is a continuous ongoing process even today.

IGES is still widely used by industries as the neutral data format for digital exchange of information among heterogeneous CAD systems. IGES provides excellent *Interoperability* by which a user can efficiently exchange drawings, circuits, diagrams and object representations in wireframe, freeform surface and solid model form. IGES has provided a solid foundation based on which the international standard STEP is being further developed.

Presenting IGES and STEP standards in details is beyond the scope of this book. In what follows, salient features of the IGES and STEP data formats will be discussed one by one

9.3.1 IGES Data Format.

IGES is a data exchange format for 2D or 3D vector image files. It can have the file extension **.igs** or **.iges**. Today all the CAD software support the export and import of CAD model files in the IGES format.

An IGES file is ASCII in nature containing information about 2D or 3D vector image file of the CAD model. A vector image comprises of lines instead of the pixels which are typically present in the raster image stored in a JPEG file.

IGES File format

An IGES file is composed of ASCII records of 80 character length, a relic of the punched card computer era. The format is quite similar to the early versions of the FORTRAN language. For example the text strings are represented by the Hollerith format like 4HPART where 4 represents the string length and PART is the string. The IGES file is human readable. IGES standard presents at length the data format and explanation of various entries therein. Interested reader can refer the relevant literature.

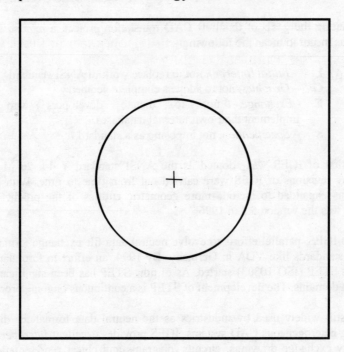

Figure 9.2 A simple CAD drawing

Figure 9.2 shows a small CAD drawing containing geometric entities such as a Point, a Circle and four Lines which make up a square. The square has diagonal points from (0, 0) to (10, 10). The circle has radius of 2.5 and centre point (5, 5). Figure 9.3 shows the IGES file representing the part geometry shown in Figure 9.2.

The IGES file is divided into 5 sections, indicated by a character in column 73 of the records. The character could be **S, G, D, P or T**. The geometric information and attributes of entities is included in two ections. One record, termed as *Directory Entry* uses fixed length format. The other record is a multiple entry, comma delimited format termed as *Parameter Data*. Figure 9.3 shows the typical records in the extracted IGES file. Geometric entities are indicated with predefined reserved words eg. POINT (Type 116), CIRCULAR ARC (Type 100), LINE (Type 110) etc. For example the POINT entity in IGES file denoted by ENTITY code 116 has a tag of D13 and its coordinates are shown in P7 as x=5, y=5. Similarly for a typical LINE denoted by code 110, (D3 – P2) the coordinates of its end points are represented as (0, 0, 0), (0,10, 0). Applications developers can write software in any programming language (C, C++) to read in such IGES files, extract relevant geometric entities and their attributes to developing downline application software for CAD/ CAM.

```
---- IGES file using analytic representation for surfaces                      S    1
1H,,1H;,12HPart4.SLDPRT,51HC:\Documents and Settings\cam lab\Desktop\2DIG G    1
ES.IGS,15H----,15H----,32,308,15,308,15,12HPart4.SG      2
LDPRT,1.,2,2HMM,50,0.125,13H101231.101800,1E-008,499990.,6Hcamlab,,11,  G     3
0,13H101231.101800;                                                     G     4
     124    1    0    0    0                                  00000000D   1
     124    0    0    1    0                                        0D    2
     110    2    0    0    0         1                        00020000D   3
     110    0    0    1    0                                        0D    4
     110    3    0    0    0         1                        00020000D   5
     110    0    0    1    0                                        0D    6
     110    4    0    0    0         1                        00020000D   7
     110    0    0    1    0                                        0D    8
     110    5    0    0    0         1                        00020000D   9
     110    0    0    1    0                                        0D   10
     100    6    0    0    0         1                        00020000D  11
     100    0    0    1    0                                        0D   12
     116    7    0    0    0         1                        00020000D  13
     116    0    0    1    0                                        0D   14
     402    8    0    0    0                                  00000000D  15
     402    0    0    1   16              2DSKETCH                  1D   16
     314    9    0    0    0                                  00000200D  17
     314    0    8    1    0                                        0D   18
124,1.,0.,0.,0.,0.,1.,0.,0.,0.,0.,1.,0.;                             1P    1
110,0.,0.,0.,10.,0.,0.;                                              3P    2
110,10.,0.,0.,10.,10.,0.;                                            5P    3
110,10.,10.,0.,0.,10.,0.;                                            7P    4
110,0.,10.,0.,0.,0.,0.;                                              9P    5
100,0.,5.,5.,7.39076189,5.730929262,7.39076189,5.730929262;        11P    6
116,5.,5.,0.,;                                                      13P    7
402,1,6,1,3,5,7,9,11,13;                                            15P    8
314,79.6078431372549,82.3529411764706,93.7254901960784,,;          17P    9
S    1G    4D   18P    9                                              T    1
```

Figure 9.3 IGES File

IGES is the most widely used neutral file format for exchange of CAD model information between heterogeneous CAD systems. All CAD software thus, support the export and import of CAD model files in IGES format.

9.4 STEP STANDARD

STEP (**St**andard for the **E**xchange of **P**roduct Model Data) is an evolving International standard for the exchange of technical and business data in electronic format. It is based on a public data model to share complex technical captive data of an enterprise internally as well as globally with its worldwide partners, customers, subcontractors and suppliers. STEP gives a very distinct advantage to digitally connect and interact with enterprises for global product development and business opportunities. Integrating CAD/CAM data using STEP have provided to some organizations, significant benefits such as reduced design development time, reduced cost, reuse of design and manufacturing data particularly for complex products developed globally. Salient features of STEP are presented here.

9.4.1 What is STEP?

STEP is an International standard (ISO 10303) for exchanging product data between different (heterogeneous) CAD/CAM and Product Data Management (PDM) systems. In today's environment, Product Data is stored in many different systems in different formats with very little integration and lot of redundancy. Engineering drawings may be in some proprietary format while information on material, process, finish, packaging, sales etc may be stored in variety of documents in some other computer format. STEP can integrate such islands of data by providing a single product data storage standard which is common, open and vendor independent.

To use STEP, an enterprise needs to have software tools which will translate in and out their CAD/CAM and PDM data from their native proprietary format into neutral format specified by STEP. Compared to similar attempts of creating neutral data standards worldwide (IGES, VDA), STEP is capable of storing all data for a product throughout its Product Life Cycle without regard to the discipline or application area.

9.4.2 Application Protocols in STEP

Application Protocol (AP) in STEP defines all data needed for a particular application domain. For being conformant with STEP, the software systems have to be able to support all the data defined by AP. This in turn, ensures that all data can be translated *in* and *out* of this internal format *without* loss. STEP APs rightly address product identification and composition as well as the information relating to Bill of Materials (BOM) and part lists.

Passing through the phases of proposal, review, revision and acceptance, several STEP APs have matured today. They are considered as viable data exchange solutions for the specific application domains. STEP APs can be broadly grouped into 3 main areas viz *Design, Manufacturing and Lifecycle support.*

Listed below are the most popular Application Protocols of interest in CAD/ CAM which have matured and are being implemented today in the Mechanical design/manufacturing related industries. For a comprehensive list, one can refer to the STEP standard.

Design APs – Mechanical
- AP 201 – Explicit Drafting
- AP 202 – Associative Drafting
- AP 203 – ConFigureuration controlled 3D design of mechanical parts and assemblies
- AP 204 – Mechanical Design using Boundary representation
- AP 207 – Sheet metal die planning and design
- AP 209 – Composite and metallic structural analysis and related design
- AP 214 – Core data for automotive mechanical design processes
- AP 235 – Materials information for the design and verification of products.

Manufacturing APs
- AP 219 – Dimensional inspection information exchange
- AP 223 – Exchange of design and manufacturing product information for cast parts
- AP 224 – Mechanical product definition for process plan using machining features.
- AP 238 – Application interpreted model for computer numeric controllers
- AP 240 – Process plans for machined products.

Life Cycle support APs
- AP 239 – Product Life cycle support
- AP 221 – Functional data and schematic representation of process plans.

Though the primary aim of STEP is to have one integrated data model for entire lifecycle support, due to the complexity of the data and requirements of application domains, several AP's with overlapping areas got created. This created some difficulties in interoperability. Efforts are on towards harmonizing the following AP's
- AP 214, 203, and 242 for 3D Mechanical design
- AP 201, 202, 212, 214 and 221 for technical drawings
- AP 214, 224 and 238 for machining features
- AP 203 (2), 210, 214, 224 and 238 for Geometric dimensioning and Tolerancing.

9.4.3 EXPRESS Language

STEP is based on Information Modeling language called as EXPRESS which gives it the important characteristics of extensibility. EXPRESS is the fundamental tool to describe the information models and application protocols which form the bulk of the standard. The Information Models and Application Protocols describe the Data structures and constraints of a complete product model and enable movement of application data between tools to make it available to developers. Data defined by EXPRESS can be manipulated by programming language like C, C^{++}.

STEP implementation enables exchange of product information across range of CAD systems, BOM systems, stand alone data translators and finally to application software. Each AP is a formal document describing the activities in Product Life Cycle. Together with EXPRESS information model, it forms an Application Interpreted Model (AIM). A STEP development environment is often written around the data dictionary to provide library of functions to *create*, *access*, *update* or *destroy* EXPRESS defined data.

EXPRESS language is computer sensible and so a compiler is provided to transform the AIM (EXPRESS model) into variety of useful forms, such as C^{++} classes or SQL data base definitions. It is not possible to include all the EXPRESS definitions here in a limited space. For comprehensive definitions of the EXPRESS constructs, the standard can be referred. EXPRESS definition format is briefly introduced here.

9.4.3.1 EXPRESS Definition Format.

Figure 9.4 shows the format of schema for EXPRESS definition of a typical AP (e.g AP 214). Individual schemas for the TYPE, ENTITY etc are also specified.

```
-- AIM EXPRESS long form for ISO/IS 10303-214:2009 (ed 3)
-- ISO TC 184/SC 4/WG 3 N2628
-- Title: Core data for automotive mechanical design processes
-- Author:
-- Date: --
-- Information object registration: Schema identification:
SCHEMA AUTOMOTIVE_DESIGN;
(* Original schemas:
  schema = action_schema ;
  schema = aic_advanced_brep ;
  ...
  CONSTANT;
   TYPE action_item;
  --------------------------------
   TYPE b_spline_curve_form;
   TYPE b_spline_surface_form;
   ...
   TYPE surface_tolerance_parameter;
  ----------------------------------------
   TYPE unit;
  --------------------------------
    ENTITY advanced_brep_shape_representation;
   ENTITY b_spline_curve;
   ENTITY b_spline_curve_with_knots;
   ...
   ENTITY edge_based_wireframe_shape_representation;
  ------------------------------------------------
   FUNCTION valid_wireframe_edge_curve;
   FUNCTION valid_wireframe_vertex_point;
  ------------------------------------------------
   RULE application_protocol_definition_required;
   RULE approval_person_organization_requires_date_time;
  ------------------------------------------------
END_SCHEMA;
```

Figure 9.4 Schema for EXPRESS

Figure 9.5 shows the Schema for defining a typical geometric entity viz B-Spline curve. Its explicit and derived attributes are also defined.

(* **SCHEMA** AUTOMOTIVE_DESIGN; *)

ENTITY b_spline_curve
SUPERTYPE OF (ONEOF (uniform_curve, b_spline_curve_with_knots, quasi_uniform_curve, bezier_curve) **ANDOR** rational_b_spline_curve)
SUBTYPE OF (bounded_curve);
  degree : **INTEGER;**
  control_points_list : **LIST** [2:?] **OF** cartesian_point;
  curve_form : b_spline_curve_form;
  closed_curve : **LOGICAL;**
  self_intersect : **LOGICAL;**
DERIVE
  upper_index_on_control_points : **INTEGER** := SIZEOF(control_points_list) - 1;
  control_points : **ARRAY** [0 : upper_index_on_control_points] **OF**
  cartesian_point := list_to_array(control_points_list, 0,
  upper_index_on_control_points);
WHERE
  wr1 : ('AUTOMOTIVE_DESIGN.UNIFORM_CURVE' **IN TYPEOF(SELF)) OR (**
    'AUTOMOTIVE_DESIGN.QUASI_UNIFORM_CURVE' **IN TYPEOF(SELF)) OR (**
    'AUTOMOTIVE_DESIGN.BEZIER_CURVE' **IN TYPEOF(SELF)) OR (**
    'AUTOMOTIVE_DESIGN.B_SPLINE_CURVE_WITH_KNOTS' **IN TYPEOF(SELF));**
END_ENTITY; -- *10303-42: geometry_schema*
Explicit Attributes
Name, Degree, Control point List, Curve form, Closed Curve, Self Intersect
Derived Attributes
Dimension, Upper Index on Control points, Control Points

Figure 9.5 Schema for B Spline Curve Entity in EXPRESS

A simple EXPRESS entity definition for POINT can be as under

```
ENTITY cartesian _point
SUBTYPE OF (point);
coordinates: LIST [1: 3] OF length _measures
END – ENTITY; _ Cartesian _ point
```

Translating into C^{++} class, a Point object can be created and filled in with its attributes as under.

```
/* Create a point using default constructor and set its values*/
SdaiModel : mod;
Point H point 1 = SdaiCreate (mod, Point);
Point 1      ⟶   x (1.0);
 Point 1      ⟶   y (0.0);
```

The Application Information Model (AIM) defined by EXPRESS is bound at its early stage of implementation to a target language like C^{++}. The compiler will type check, support inheritance and allow methods to be attached to classes. Each class has *access* and *update* methods for simple derived attributes and constructors to initialize new instances. EXPRESS language has the ability to capture rules and constraints. These are the most formal representations of the original design intent of the AP developers and can be used to check up the data being produced.

The first task in STEP implementation is to determine appropriate AP for the problem domain. Using the chosen AP, the Application Information Model (AIM) is developed using the EXPRESS language. STEP implementation software tools are available from developers or could be developed in – house which convert the AIM into STEP files. These are exchanged with test sites to detect any semantic problems. As project develops, EXPRESS Rules and Constraints can be introduced. All along the process, conformance tests can be carried out to check the semantics of the AP and the syntax of the exchange files. Today most of the solid modeling systems provide a facility of exporting CAD model in the STEP format. The format is explained below.

9.4.3.2 STEP File Format.

The format of a STEP file is defined in ISO - 10303- 21. The file extensions **.stp** or **.step** indicates that the file contains data conforming to STEP Application Protocol. STEP file is ASCII and easy to read with typically one instance per line. Figure 9.6 shows the typical format of the STEP file. Following the keyword ISO - 10303- 21, the file is split into two sections – HEADER and DATA.

ISO - 10303- 21
HEADER section
- FILE DESCRIPTION
- *description*
- *implementation_level*
- FILE NAME
- *name* of exchange structure
- *time _ stamp*
- *author*
- *organization*
- *preprocessor*
- *originating _ system*
- *authorization*
- FILE POPULATION
- SECTION LANGUAGE
- SECTION CONTEXT

DATA Section
- *Instance name*
- *Instances of single entry data types*
- *Instances of complex entity data types*
- *Mapping of attribute values*
- *Mapping of the other data types*

Figure 9.6 STEP File Format

Figure 9.7 shows an abridged version of an extracted STEP file representation of the 2D CAD model shown in Figure 9.2. The file is verbose and so some intermediate data records are omitted here for the purpose of illustration.

```
ISO-10303-21;
HEADER;
FILE_DESCRIPTION (( 'STEP AP203' ),'1' );
FILE_NAME ('2Dstep.STEP', '2010-12-31T06:42:15',('camlab'),('iitb'), 'SwSTEP 2.0','********' " );
FILE_SCHEMA (( 'CONFIGURE_CONTROL_DESIGN' ));
ENDSEC;

DATA;
#1 =( NAMED_UNIT ( * ) SI_UNIT ( $, .STERADIAN. ) SOLID_ANGLE_UNIT ( ) );
#2 = UNCERTAINTY_MEASURE_WITH_UNIT (LENGTH_MEASURE(
1.00000000000000100E-005 ), #35, 'distance_accuracy_value', 'NONE');
#3 = GEOMETRICALLY_BOUNDED_WIREFRAME_SHAPE_REPRESENTATION ( '2Dstep', (
#16, #7 ), #4 ) ;
#4 =( GEOMETRIC_REPRESENTATION_CONTEXT ( 3 )
GLOBAL_UNCERTAINTY_ASSIGNED_CONTEXT ( ( #2 ) )
GLOBAL_UNIT_ASSIGNED_CONTEXT ( ( #35, #34, #1 ) ) REPRESENTATION_CONTEXT (
'NONE', 'WORKASPACE' ) );
#5 = DIRECTION ('NONE',  ( 1.000, 0.000, 0.000 ) ) ;
#6 = DIRECTION ( 'NONE',  ( 0.000, 0.000, 1.000 ) ) ;
#7 = AXIS2_PLACEMENT_3D ( 'NONE', #8, #6, #5 ) ;
#8 = CARTESIAN_POINT ( 'NONE',  ( 0.000, 0.000, 0.000 ) ) ;
#9 = DIRECTION ( 'NONE',  ( -1.00, 0.000, 0.000 ) ) ;
#10 = CARTESIAN_POINT ( 'NONE',  ( 10.000, 10.000, 0.000 ) ) ;
#11 = TRIMMED_CURVE ( 'NONE', #12, ( PARAMETER_VALUE ( 0.000 ), #23 ), (
PARAMETER_VALUE ( 1.000 ), #22 ), .T., .PARAMETER. ) ;
#12 = LINE ( 'NONE', #10, #24 ) ;
#13 = CARTESIAN_POINT ( 'NONE',  ( 10.000, 10.000, 0.000 ) ) ;
#14 = CARTESIAN_POINT ( 'NONE',  ( 10.000, 0.000, 0.000 ) ) ;
-------------------------- ---------- ------------------- ----------------------
#23 = CARTESIAN_POINT ( 'NONE',  ( 10.000, 10.000, 0.000 ) ) ;
#24 = VECTOR ( 'NONE', #9, 10.000) ;
#25 = TRIMMED_CURVE ( 'NONE', #40, ( PARAMETER_VALUE ( 0.000 ), #37 ), (
PARAMETER_VALUE ( 0.999 ), #36 ), .T., .PARAMETER. ) ;
#26 = DIRECTION ( 'NONE',  ( 0.956, 0.292, 0.000)) ;
#27 = DIRECTION ( 'NONE',  ( 0.000, 0.000, 1.000)) ;

#28 = CARTESIAN_POINT ( 'NONE', ( 5.000, 5.000, 0.000) ) ;
#29 = AXIS2_PLACEMENT_3D ( 'NONE', #28, #27, #26 ) ;
#30 = CIRCLE ( 'NONE', #29, 2.499 ) ;
------------------------ ----------------------- ------------------------
#45 = PRODUCT ( '2Dstep', '2Dstep', '', ( #46 ) ) ;
```

Figure 9.7 Extract of a STEP file (Contd.)

```
#46 = MECHANICAL_CONTEXT ( 'NONE', #48, 'mechanical' ) ;
#47 = APPLICATION_PROTOCOL_DEFINITION ( 'international standard',
'conFigure_control_design', 1994, #48 ) ;
#48 = APPLICATION_CONTEXT ( 'conFigureuration controlled 3d designs of mechanical parts and
assemblies' ) ;
#51 = APPROVAL ( #80, 'UNSPECIFIED' ) ;
#53 = LOCAL_TIME ( 12, 12, 15.000, #54 ) ;
-------------------- ---------------------- ---------------------- ----------------------
#62 = DESIGN_CONTEXT ( 'detailed design', #60, 'design' ) ;
#63 = APPLICATION_PROTOCOL_DEFINITION ('international standard',
'conFigure_control_design', 1994, #60 ) ;
-------------------- ---------------------- ---------------------- ----------------------
#66 = PERSON_AND_ORGANIZATION_ROLE ('creator' ) ;
#67 = PERSON_AND_ORGANIZATION ( #50, #49 ) ;
--------------------- -------------------------- -----------------------
#107 = APPROVAL_PERSON_ORGANIZATION ( #112, #109, #108 ) ;
#108 = APPROVAL_ROLE ( '' ) ;
--------------------- -------------------------- -----------------------
#112 = PERSON_AND_ORGANIZATION ( #50, #49 ) ;
#113 = CC_DESIGN_DATE_AND_TIME_ASSIGNMENT ( #101, #114, ( #118 ) ) ;
#114 = DATE_TIME_ROLE ( 'classification_date' ) ;
#115 = CALENDAR_DATE ( 2010, 31, 12 ) ;
--------------------- -------------------------- -----------------------
#118 = SECURITY_CLASSIFICATION ( '', '', #119 ) ;
#119 = SECURITY_CLASSIFICATION_LEVEL ( 'unclassified' ) ;
--------------------- -------------------------- -----------------------
ENDSEC;
END-ISO-10303-21;
```

Figure 9.7 Extract of a STEP file

It is particularly important to see the format of representation of the geometric ENTITIES. For example Point is represented as CARTESIAN POINT with attributes of *Name* and its coordinate tuple (x, y, z). Similarly a LINE is represented in terms of attributes like its *Start Point* and *Vector*. Application developers can write software in any programming language (C, C^{++}) to read in such STEP files and extract geometric (other) entities and their attributes to develop downline application software in CAD / CAM. The complete file format is explained at length in the STEP standard.

9.4.4 Strengths of STEP

STEP is an international standard that is expected to replace multiple, fragmented standards and proprietary data formats to ensure fast and reliable data exchange among distributed heterogeneous CAD / CAM / PDM systems. Key strengths of STEP are listed below

- Neutral data exchange between dissimilar systems, both in – house as well as with supply chain partners, (product developers, suppliers, customers).
- Enterprise integration by paperless product definition.

- Life Cycle support for Product development revision and maintenance
- Long term archiving due to the system independent STEP architecture.
- Concurrent and/ or Collaborative Engineering
- Internet based communication and exchange of product data for global product development/ business.

9.5 FUTURE TRENDS

Stiff global competition is forcing industries worldwide to look for newer solutions for enterprise integration and global product development. Efficient product data exchange and *Interoperability* are the key requirements for the sustenance and growth of such global enterprises.

During the last decade, Internet technology, popularly termed as World Wide Web (www) has been developed to enable distributed computing. Web provides excellent interoperability to integrate globally distributed environments for efficient transfer/ exchange of data. As a result, several web applications are developed and implemented worldwide for the exchange of information in the form of text, video and sound. Most of these web based applications focus on e- commerce.

Realizing the benefits of global interoperability, researchers and developers worldwide are trying to develop Internet based CAD/CAM systems. These essentially provide network centric Collaborative environment wherein globally distributed clients (users) can simultaneously interact with each other through the central server to enable collaborative Product Development. Such systems are perceived to provide benefits like significant reduction in the product development time, tapping of global resources/expertise and possibility of no rework during design/ manufacturing stages. Internet based Product Development however, throws a new set of challenges for the representation and efficient transmission of product data across geographically distributed clients. The product model schema needs to be very light to suit the limited bandwidth of the net. Evolving efficient product data standard is thus, a key issue in internet based collaborative CAD/CAM scenario..

As of now, STEP is the most comprehensive, evolving international standard which supports engineering, manufacturing, electrical/ electronics, architecture and construction Life Cycle information. Several forward looking industries belonging to automotive, aerospace, shipbuilding, electronics and the process industry sectors around the world are thus, actively implementing STEP for global enterprise integration and business operations. With continuous development in progress, STEP will soon become the de facto international standard for Product Data Exchange.

Bibliography

1. Bedworth David, Henderson Mark and Wolfe Philip, *Computer Integrated Design and Manufacturing*, McGraw Hill International Book Company, NY, 1991
2. Lee Kunwoo, *Principles of CAD/CAM/CAE Systems*, Addison Wesley 1999
3. Foley J.D., A Van Dam and S.K. Feiner, *Fundamentals of Interactive Computer Graphics*, Addison-Wesley, NY, 1993
4. Faux I. D. and Pratt M. J., *Computational Geometry for Design and Manufacture*, Ellis Horwood Ltd. 1981
5. Farin G, *Curves and Surfaces for Computer Aided Geometric Design*, Academic Press, New York, 2001.
6. Rogers D.F. and Adams J.A., *Mathematical Elements for Computer Graphics*, McGraw Hill International Book Company, 1990
7. Mortenson M. E., *Mathematics for Computer Graphics Applications*, Industrial Press Inc, New York, 1999
8. Bowyer A. and Woodwark J., *Introduction to Computing with Geometry*, Information Geometers Ltd, UK, 1993
9. Rogers D.F. , *Procedural Elements for Computer Graphics*, McGraw Hill Book Company, 1985
10. Newman W. M. and Sproull R. F., *Principles of Interactive Computer Graphics*, McGraw Hill Book Company, 1983
11. Preparata F. P. and Shamos M. I., *Computational Geometry : An introduction*, Springer Verlog, 1985
12. Mortenson M. E., *Computer Graphics Handbook, - Geometry and Mathematics*, Industrial Press Inc, 1990
13. Choi B. K., *Surface Modeling for CAD/CAM*, Elsevier, 1991
14. Woodward C.D, Cross sectional design of B-spline surfaces, *Computers and Graphics,* Pergamon, v11, n2, 1987, pp. 193-201.
15. Piegl L. and Tiller W., Curve and Surface Constructions using Rational B-splines, *Computer Aided Design*, V19, n9, 1987, pp.485-498
16. Barsky B. A., *Computer Graphics and Geometric Modeling Using Beta Splines*, Springer Verlag, 1988
17. Gurunathan B. and Dhande S.G. Algorithms for development of certain classes of ruled surfaces, *Computers and Graphics*, Pergamon, v11,n2,1987,pp.105-112
18. Bezier P., Mathematic Basis of the UNISURF CAD System, *Butterworths*, 1986
19. Sunil V. B. and Pande S. S. , Automatic Recognition of Features from Freeform Surface CAD models, *Computer Aided Design*, Butterworth, v 40, issue 4, 2008, pp 502-517.
20. Requicha A.A.G., Representation for Rigid Solids: Theory, Methods and Systems, *Computing Surveys*, v12, 1980, pp. 437-464
21. Hoffman C., *Geometric and Solid Modeling*, Morgan Kanfman, 1989
22. Pickett Mary and Boyse John (Ed), *Solid Modeling by Computers : From Theory to Applications*, Plenum Press, NY, 1984
23. Mortenson M.E., *Geometric Modeling*, Industrial Press, NY, 2006

24. Jaques M.W.S, Billingsley J. and Harrison D., Generative feature based design-by-constraints as a means of integration within the manufacturing industry, *Computer Aided Engg Journal*, IEE Publication, Dec. 1981, pp. 261-267

25. Gao S., Wan H. and Peng Q., Constrain Based Virtual Solid Modeling, *Journal of Computer Science and Technology*, v15, n1, 2000 pp. 56-63

26. Shah Jami J. and Mantyala M., *Parametric and Feature Based CAD/CAM*, John Wiley & Sons, Inc, 1995

27. Prabhu B.S. and Pande S.S., Intelligent Interpretation of CADD Drawings, *Computers and Graphics*, v23, n1, 1999, pp. 25-44

28. Joshi S. and Chang T.C., Graph Based Heuristics for Recognition of Mechanical Features from 3D Solid Model, *Computer Aided Design*, v30, n2, 1988, pp. 58-66

29. Kulkarni V.S. and Pande S.S, A System for Automatic Extraction of 3D Part Feature Using Syntactic Pattern Recognition Techniques, *International Journal of Production Research*, Taylor and Francis, v33, n6, 1995, pp. 1569-1586

30. Sunil V. B. , Agrawal Rupal and Pande S. S., An Approach to recognize interacting features from B-Rep CAD models of prismatic machined parts using a hybrid (graph and rule based) technique, *Computers in Industry*, Elsevier, v 61, 2010, pp 686-701.

31. Sunil V.B. and Pande S.S., Automatic Recognition of Machining Features Using Artificial Neural Networks, *International Journal of Advanced Manufacturing Technology*, v41, issue 9-10, 2009, pp. 932-947

32. Au C.K. and Yuen M.M.F. A Feature Modeler for Sculptured Object Modeling, *Engineering with Computers*, v19, n1, 2003, pp. 1-8

33. Wu S.H., Lee K.S. and Fuh J.Y.H., Feature Based Parametric Design of a Gating System Die-casting die, *International Journal of Advanced Manufacturing Technology*, v19, 2002, pp. 821-829

34. Atul Thakur and S.S. Pande, SIBAM – A Web based Feature modeler for sheet metal components, *Proc. of the 22nd Intl. Conference on CAD/CAM, Robotics and Factories of the Future(CARS & FOF)*, V.I.T Vellore, India, 19-22 July 2006, pp. 221-227

35. Gao J.X., Tang Y.S. and Sharma R.A., A Feature Model Editor and Process Planning System for Sheet Metal Products, *Journal of Material Processing Technology*, v107, 2000, pp. 88-95

36. Desai V.S. and Pande S.S., GFM-An Interactive Feature Modeler for CAPP of Rotational Components*, Computer Aided Engineering Journal*,v5,n8,1991, pp. 217-220

37. Rembold U. Nnaji B. O. and Storr A., *Computer Integrated Manufacturing and Engineering*, Addison Wesley, 1993

38. Amaitik S.M. and Kilic S.E., STEP Based Feature Modeler for Computer Aided Process Planning, *International Journal of Production Research*, Taylor and Francis, v43, n15, 2005, pp. 3087-3101

39. Patil Lalit and Pande S.S., An Intelligent Feature Based Process Planning System for Prismatic Parts, *International Journal of Production Research*, Taylor and Francis, v40, n17, 2002, pp. 4431-4447

40. Chang T. C., Wysk R. A. and Wang H.P. , *Computer Aided Manufacturing*, Prentice Hall 2006

41. Srinivasan V., An Integration Framework for Product Lifecycle Management, *Computer Aided Design*, Butterworth, v33, issue 5, 2011, pp. 464-478

42. Danesi F., Gardan N., Gardan Yand Reimeringer M, P4LM: A methodology for Product Lifecycle Management, *Computers in Industry*, Elsevier, v59, 2008, pp. 304-317

43. Corney J, Hayes C, Sundarajan V and Wright P, The CAD/CAM Interface: a 25 year retrospective, *Journal of Computing and Information Science in Engineering*, v 5, issue 3, 2005, pp. 188-197

44. Chen X., Li M. and Gao S., A web Service for Exchanging Procedural CAD Models Between Heterogeneous CAD Systems, *Computer Supported Co-operative Work in Design, Lecture Notes in Computer Science*, Springer, v3865, 2006, pp. 225-234

45. Amar Gaonkar, Sushrut Pavanaskar and S.S. Pande, Web based Feature modeling and process planning system for CNC turning, *Proc. of 2nd Intl and 23rd All India Mfg. Technology, Design and Research Conference*, IIT Madras, India, 15-17 Dec.2008

46. Naggy M.S. and Matyasi G, Analysis of STL Files, *Mathematical and Computer Modeling*. Pergamon, v38 issue 7-9, 2003, pp. 945-960

47. Shaharoun A.M., Razak J.A. and Alam M.R., A STEP Based Geometrical Representation as Part of Product Data Model of a Plastic Part, *Journal of Material Processing Technology*, Elsevier, v76, 1998, pp. 115-119

48. *STEP Application Handbook*, ISO 10303, Version 3, Prepared by SCRA, North Charleston, SC 29418, USA, June 2006

49. S.S.Pavanaskar, A.D. Takur, V.B. Sunil and S.S. Pande, WebNC – An Internet based system for global product development, *Proc. Of the 7th C I R P global conference on Sustainable Manufacturing*, I.I.T Madras, India, 2-4 Dec.2009

50. Wikipedia, http://en.wikipedia.org/wiki/Computer-aided_design, May 18, 2011

Index